FOCUS ON NANOTUBE RESEARCH

FOCUS ON NANOTUBE RESEARCH

DELORES A. MARTIN
EDITOR

Nova Science Publishers, Inc.
New York

For permission to use material from this book please contact us:
Telephone 631-231-7269; Fax 631-231-8175
Web Site: http://www.novapublishers.com

NOTICE TO THE READER

LIBRARY OF CONGRESS CATALOGING-IN-PUBLICATION DATA
Focus on Nanotube research / Delores A. Martin (editor).
p. cm.
Includes index.
ISBN 1-59454-851-X
1. Nanotubes. I. Martin, Delores A.
TA418.9.N35F65 2006
620'.5--dc22 2005031834

Published by Nova Science Publishers, Inc. ◻ New York

CONTENTS

PREFACE

A novel way of arranging the atomic structure of a substance so that it can be made thousands of times stronger than in its native state. Often used to make duranium a further ten thousand times stronger. Thus, a lump of duranium can be made over ten million times stronger than the equivalent block of titanium. A one dimensional fullerene (a convex cage of atoms with only hexagonal and/ or pentagonal faces) with a cylindrical shape. Carbon nanotubes discovered in 1991 by Sumio Iijima resemble rolled up graphite, although they can not really be made that way. Depending on the direction that the tubes appear to have been rolled (quantified by the 'chiral vector'), they are known to act as conductors or semiconductors. Nanotubes are a proving to be useful as molecular components for nanotechnology. This new book assembles and presents new and important research in the field.

As explained in Chapter 1, functionalization methods for single-wall carbon nanotubes have been developed, and through such techniques a variety of intermolecular and interface interactions can be envisaged. Moreover, the incorporation of functionalized SWNTs into composite matrices has been achieved and substantial mechanical reinforcement demonstrated. The studies of carbon nanotubes by fluorescence spectroscopy have contributed into clarifying that there are no fluctuations in the radiation intensity coming from the isolated nanotubes, which makes them as potentially infrared stable photon sources with a very narrow bandwidth.

Chapter 2 reviews an electronic pairing mechanism driven by correlation, i.e. by the on-site repulsion; it works in a Hubbard model with the symmetry of the wrapped honeycomb lattice, suitable for single-wall *(N; N)* carbon nanotubes. There is an important analogy with the one-band and three-bands Hubbard model for Cuprates: the Hubbard Hamiltonian for nanotubes also admits two-body singlet eigenstates with vanishing on-site repulsion, which we call $W = 0$ pairs. By means of a non-perturbative canonical transformation the authors calculate the effective interaction between the electrons of a $W = 0$ pair added to the interacting ground state. The interesting possibility of bound states is borne out by the approach provided the nanotube is doped away from half filling. Then, the *dressed* $W = 0$ pair in the interacting many-body system is a bound state, and this occurs with quite reasonable parameter values. The pairing mechanism uses electronic degeneracy (or quasi-degeneracy) to produce attraction from repulsion. Pairing in Hubbard models has been reported by other authors as well, but the present approach has certain specific advantages. For example, it shows how pairing is favored by a large noncommutative symmetry Group (the Space Group, possibly supplemented by extra operations). This finding strongly suggests that the recently discovered superconductivity in carbon nanotubes could be at least partially due to an electronic mechanism; the physics involved could have some features in common with the case of High T_c compounds. From the effective interaction in the nanotubes they compute pair binding energies; they use exact diagonalization results for the (1,1) nanotube to further confirm the analytic theory, then they consider the general trend. There is an optimal doping when the number of electrons per C atom is in the range 1.2÷1.3. For *(N; N)* nanotubes of length *L*, the binding energy of the pair depends strongly on the filling and decreases towards a limiting value of the order of 0.1 ÷ 1 meV as L $\rightarrow \infty$. The authors also present preliminary results on the inclusion of phonons in the present theory of pairing in the Hubbard model. Due to the need to include several oscillator branches the problem is hard and they resort to small cluster calculations, to start forming broad criteria. The results indicate that a strong Jahn-Teller effect induced by vector phonons may destroy the symmetry; the correlation-induced pairing, which is driven by the symmetry, is broken in such cases. However they also show examples where electronic and phonon-induced pairing are synergic, both at weak coupling and in the polaronic regime, and the distortion is dynamically avoided.

In Chapter 3, the authors discuss the possibility of the existence of single-wall carbon nanotubes (SWNTs) in organic solvents in the form of clusters, containing a number of SWNTs. A theory is developed based on a bundlet model for clusters, which enables describing the distribution function of clusters by size. Comparison of the calculated values of solubility with experimental data would permit obtaining energetic parameters characterizing the interaction of an SWNT with its surrounding, in a solid phase or solution. Fullerenes and SWNTs are unique objects, whose behaviour in many physical situations is characterized by remarkable peculiarities. Peculiarities in solutions show up first in that fullerenes and SWNTs represent the only soluble forms of carbon. This is primary related to the originality in the molecular structure of fullerenes and SWNTs. The fullerene molecule is a virtually uniform closed spherical or spheroidal surface, having no sharp ridges or dents. Similarly, an SWNT is a smooth cylindrical unit. Both structures give rise to the relatively weak interaction between the neighbouring molecules in a crystal, and promote effective interaction of the molecules with those of a solvent. Another peculiarity in solutions is related to their tendency to form clusters, consisting of a number of fullerene molecules or SWNTs. The energy of interaction of a fullerene molecule or SWNT with solvent molecules is proportional to the surface of the former molecule, and roughly independent of the relative orientation of solvent molecules. All these phenomena have a unified explanation in the framework of the bundlet model of a cluster, in accordance with which the free energy of an SWNT involved in a cluster is combined from two components, *viz*. a volume one, which is proportional to the number of molecules n in a cluster, and a surface one, which is proportional to $n^{1/2}$. Both single- and complete-linkage cluster analyses (CAs) of the packing properties of equal objects allow classifying these properties and agree. Principal components analysis of the packing properties and CA of the packing objects permit classifying them and agree. The packing objects, clusters and other systems of atoms with short-range interaction, are grouped into three classes. The bundlet model for clusters enables describing the distribution function of SWNT clusters by size. A droplet model for clusters of conical single-wall carbon nanohorns (SWNHs) is also proposed. The droplet model is formally analogous to that for clusters of fullerenes. From purely geometrical differences the models predict different behaviours.

In Chapter 4, the authors report their efforts in developing laser beams as simple means to fabricate patterned arrays of aligned CNTs or three-dimensional structures made of CNTs without a pre-patterned substrate. Two related techniques were developed. The first technique makes use of a tightly focused laser beam from a moderate power He-Ne laser as precision trimming tool to create unique 3D CNTs structures. The details of the technique will be presented. The technique can be used to modify the surface morphology of the CNTs forest. The technique also allows for a wider range of three-dimensional CNTs microstructures to be constructed. In addition, some of the structures created were found to exhibit laser-beam induced mechanical motion. Such behavior can be exploited in creating possible micro-optomechanical devices made of CNTs. They demonstrated the feasibility of a simple optomechanical switch with this method. The second technique makes use of a parallel or converging laser beam from a high power UV pulsed laser for the destruction of aligned CNTs with a copper grid as lithography mask to define patterned aligned CNTs arrays. This technique is simple to implement, inherently parallel and has high throughput. It is well suited for creating a large extended area of patterned CNTs array.

It is shown in chapter 5 that Carbon nanotubes are undoubted one of the most interesting materials in the last twenty years. The outstanding properties that these materials possess have opened a new interesting research area in nanoscience and nanotechnology, in addition, new properties have been found and possible applications are suggested for these materials frequently. Nowadays the research in nanotubes covers different fields focus to understand and applied the interesting properties of these nanomaterials, and other part have been directed to basic science with the target to modify structure and surface of carbon nanotubes. This later has allowed diversify their properties and possible uses, opening new possibilities in chemistry, physics, engineering, medicine and materials science. One of the most mentioned applications for these carbon materials is their incorporation as reinforcement in polymer matrices in order to produce new nanocomposites, taking advantage of the amazing structural, mechanical, electrical, chemical and thermal properties that carbon nanotube possess. In this context carbon nanotube nanocomposites have been developed by different research groups with the aim to produce new ultra-strong and light materials, in addition electrical conductivity and thermal properties have been studied for this kind of nanocomposites. Nevertheless, a new challenge that allows making a new age of multifunctional composite materials with these nanometric forms is starting and therefore the study of new properties in these nanocomposites is increasing significantly by scientific groups in the last five years. In this chapter the authors present different results of mechanical, thermal and electrical properties in carbon nanotube polymer composites, with special emphasis in novel anisotropic tribological and conductive properties that have been found recently in these nanocomposites. Chemical functionalization with focus to improve compatibility between these nanoforms and polymers is also discussed. In addition are presented the results obtained with polymer nanocomposites developments with modified nanotubes. The relevance of understand the behavior on the interface between nanotubes and polymer matrices to control and improve different properties and this way reach outstanding and novel multifunctional applications with this materials is discussed continuously.

In: Focus on Nanotube Research
Editor: Delores A. Martin, pp. 1-18

ISBN: 1-59454-851-X

Chapter 1

CONFINED TUBULAR CARBON NANOSTRUCTURES FOR NANOREACTORS AND SYNTHONS

V. N. Khabashesku[1*], E. V. Barrera[2], and R. F. M. Lobo[3**]

[1]Department of Chemistry and Center for Nanoscale Science and Technology, Rice University, 6100 Main street, Houston, TX, 77005-1892, USA

[2]Department of Mechanical Engineering and Materials Science, Rice University, 6100 Main street, Houston, TX, 77005-1892, USA

[3]Grupo de Nanotecnologia e Ciência à Nano-Escala, Universidade Nova de Lisboa, 2829-516 Caparica, Portugal

Abstract

Functionalization methods for single-wall carbon nanotubes have been developed, and through such techniques a variety of intermolecular and interface interactions can be envisaged. Moreover, the incorporation of functionalized SWNTs into composite matrices has been achieved and substantial mechanical reinforcement demonstrated. The studies of carbon nanotubes by fluorescence spectroscopy have contributed into clarifying that there are no fluctuations in the radiation intensity coming from the isolated nanotubes, which makes them as potentially infrared stable photon sources with a very narrow bandwidth.

1 Introduction

Unlike other refractory elements, the vapor of carbon at temperatures in the 3000-4000 K range is dominated by clusters C_n with substantial abundance of species as large as C_{15} being in equilibrium with the solid phase. The first evidence of this goes back to early research on nuclear fission products. It is well known that carbon clusters of up to 10 atoms exist in the form of linear chains, and they can adopt the form of monocyclic rings above that number. Even though clusters of most metals also adopt compact structures, they still do not have a sufficiently high cohesive energy to be abundant in the equilibrium vapor. Instead, the vapor

* E-mail address: khval@rice.edu
** E-mail address: rfl@fct.unl.pt

of these metals is almost completely monoatomic. Above 1000 K the vapor of every element, except carbon, is predominantly either monoatomic or diatomic in equilibrium state with the pure condensed phase. Once molecular beams of refractory materials became available, a vast new area of research on clusters was opened, where each cluster may be considered as a nanoscale crystalline particle that has a surface (in fact, most of it is surface).

Tubular nanocrystalline carbon clusters (carbon nanotubes) are structurally built from rolled graphene sheets closed at the ends by the fullerene caps. Single-walled nanotubes (SWNTs) consist of a single graphitic cylinder, while multi-walled nanotubes (MWNTs) are made of concentric graphitic layers surrounding the central tubule. Since the discovery of MWNTs by Ijima in 1991 [1] and preparation of SWNTs in 1993 [2-4] the carbon nanotubes became an area of intense research activity worldwide. The unique properties exhibited by carbon nanotubes offer many opportunities for their applications [5]. The band structure of graphene, the individual flat sheet of graphite, is that of a zero bandgap semi-conductor in which the valence band and conduction band meet at zero point of the end of the Brillouin zone. There is a node in the density of states at Fermi energy, and accordingly, this structure is not a very good conductor (like lead). Even after formal replacement of every carbon atom in the hexagonal lattice with Au atom, the band structure will still appear the same. It turns that this is the symmetry of the hexagonal lattice that creates the conductivity problem and not the delocalization of the π electrons on carbon. By curling the graphene sheet to form the nanotube, the symmetry that has been an obstacle in preventing metallic behavior from the flat lattice is now converted into an ally. Due to the nanotube symmetry the two bands cross at the Fermi energy of about 2/3 the way across the Brillouin Zone. Furthermore, the cohesive rigidity of the π-bond framework of the graphene sheet prevents the metallic π electrons from engendering a Peierls instability that usually plagues such one-dimensional conductors.

The addition of fluorine and other functional groups to the side walls of SWNTs surface significantly alter their physical properties (e.g., conductivity, solubility, etc.) and dramatically enhances their chemical reactivity. Therefore, the fluoronanotubes and also the oxidized nanotubes are becoming widely used as precursors for subsequent derivatizations, thus serving as chemical synthons for new functional nanoscale materials [6-8].

The extraordinary tensile strength of lightweight carbon nanotubes makes them attractive reinforcing building structures for the design of novel polymer composite materials with superior mechanical properties [4,5]. However, due to the atomically smooth non-reactive surface of nanotubes, lack of interfacial bonding inhibits load transfer from the matrix to nanotubes across the nanotube/polymer interface. These problems can be overcome by using the functionalized nanotubes that can provide multiple bonding sites to the organic/inorganic polymer matrix so that the load can be transferred to nanotubes through the covalently interlinked surfaces of polymer and nanotubes [9]. The organic functional groups, chemically attached to the nanotubes, assist effectively in unroping the nanotube bundles and improving the solubility, dispersion, processing and compatibility of carbon nanotubes when blended with the particular matrix.

There is also a great interest in using the nanotubes for medical and biological applications [10,11]. This interest is prompted by the cylindrical surface morphology of the nanotubes which are stable with respect to cage opening under *in vivo* environments. For these applications, surface-functionalization will aid the carbon nanotubes in becoming biocompatible, improving their solubility in physiological solutions and selective binding to bio-targets.

Driven largely by great potential for applications, the research on functionalization chemistry of nanotubes is also fascinating from a fundamental point of view since it can provide an in-depth knowledge of chemical reactivity in relation to the nanoscale size and shape of confined structure.

2 Functionalization of Carbon Nanotubes

Nanotubes tend to self-assemble into bundles in which up to several hundred tubes are held together by van der Waals bonding. For many applications the separation of individual nanotubes from their bundles is becoming essential. To solve this problem, two main approaches based on non-covalent [12-17] and covalent [6,7,11,18,19] modifications of nanotubes are being pursued. The covalent modifications involve chemical functionalization by the attachment of various functional groups to the ends or sidewalls of the nanotubes through covalent bonds. Besides improving the solubility and processibility, this functionalization also opens an opportunity for altering the structural and electronic properties of the nanotubes and obtaining new types of nanotube-based materials with useful properties. However, the chemical functionalizations of the nanotube ends bring only a highly localized transformation of the nanotube electronic structure and do not change the bulk properties of these materials.

The challenges faced in the modifications of the SWNT sidewalls are related to their poor reactivity caused by a much lower curvature of nanotube walls than in more reactive fullerenes [5], and to the growing strain in the tubular structure with the number and size of functional groups attached to graphene walls. The sp^2-bonding states of all carbon atoms, building the nanotube framework, facilitate the predominant occurance of addition reactions. The sidewall functionalization naturally results in a significant modification of the intrinsic properties of the nanotubes.

It was shown that the caps on MWNTs can be open by oxidation with oxygen, air or carbon dioxide at temperatures above 700 °C. Oxidation in air or oxygen results in etching away of the tube caps and the thinning of tubes through layer-by-layer peeling of the outer layers, starting from cap regions. Carbon dioxide reacts as a milder oxidizing agent which erodes and strips the cap of the tube exposing the terminated cylinder layers [20]. Heating MWNTs with an oxide Pb_3O_4 in an inert atmosphere facilitated opening at a lower temperature (~400 °C) [21].

Treatments of SWNTs by sonication in HNO_3 or H_2SO_4/HNO_3, H_2SO_4/H_2O_2 mixtures [22-24] were found to cut the nanotubes by oxidizing/etching action to a lengths ranging from 100 nm to 1 micron and generate carboxyl groups and other oxo-functionalities predominantly on nanotube open ends, according to IR and ^{13}C NMR studies [25,26]. The detailed studies have clearly demonstrated that in air the smaller diameter SWNTs are oxidized more rapidly than the larger diameter nanotubes [27,28]. The increased reactivity of a smaller diameter SWNTs was also observed in experiments using the room temperature oxidation by ozone [29].

The reactivity of the carboxylic acid groups (-COOH) at the open ends of the SWNTs have been utilized in further chemical modifications of nanotubes. The first derivatization of the open ends of the shortened SWNTs was done by Smalley et al [22] by converting the carboxylic groups to the corresponding acid chloride using $SOCl_2$. Subsequent treatment of the such obtained SWNT-COCl with NH_2-$(CH_2)_{11}$-SH yielded an amide linkage of the nanotube end to the alkanethiol. The formation of a free thiol moiety has been verified by tethering the end of the nanotube to 10 nm diameter colloidal gold particles and imaging by AFM. The alike thiol derivatized SWNTs have been assembled on gold substrates in orientation normal to the surface providing an ordered arrays of SWNTs useful for various applications, e. g., in field emission devices [30].

The direct fluorination of the SWNTs with elementary fluorine results in the sidewall-fluorinated SWNTs (fluoronanotubes). They represent a new family of nanoscale fluorocarbon materials which have been prepared by fluorination of both laser ablation of graphite grown carbon nanotubes (L-SWNTs) [23,31] and those produced by a high-pressure CO disproportionation process (HiPco-SWNTs) [32,33] using the technology developed earlier for the fluorination of graphite [34]. Extensive fluorination studies [35-37] were carried out to establish optimal conditions (reaction temperatures, reaction times, addition of HF catalyst) to reach a saturation stoichiometry (nearly C_2F) without destruction of the tube structure.

The SWNTs are essentially all destroyed (i. e., "unzipped") when fluorinated at 400 °C and above to form a fluorographite and some multi-wall carbon nanotube-like material. As a result of the side-wall functionalization of the SWNTs by fluorine the electrical properties of the fluoronanotubes differ dramatically from those of pristine SWNTs. The fluoronanotubes, prepared by fluorination at temperatures of 250 °C and above, are insulators (two-point resistance across the length of buckypaper sample > 20 MΩ), while the pristine nanotubes are good conductors (two-point resistance 10-15 Ω) [36].

The HiPco-SWNTs have a smaller average diameters than L-SWNTs and, therefore, due to a higher curvature are more reactive. This is particularly indicated by the observation that under the same fluorination conditions more fluorine can be attached onto the sidewalls of HiPCO-SWNTs [37]. For instance, in the presence of HF the near C_2F composition for the tubular structure of the HiPco-SWNTs has been produced at a fluorination temperature as low as 150 °C, while under the same conditions the L-SWNTs yielded the fluoronanotubes with a significantly lower fluorine content (C/F ratio higher than 3). EELS calculations using density functional theory confirmed the existence of ionic and covalent bonding of fluorine to carbon in these fluoronanotubes [38].

The fluoronanotubes prepared from SWNTs form metastable solutions in DMF, THF, and alcohols after sonication. The solvation of individual ("unroped") fluoronanotubes was verified by dispersing them on a mica substrate and examining with atomic force microscopy (AFM) [39]. The solutions in alcohols were stable for a few days to over a week. Among the series of alcohols studied, 2-propanol and 2-butanol were found to be the best solvents. Such solvation has been explained by hydrogen bonding between the hydroxyl hydrogen in alcohol and the nanotube-bound fluorine: R-O-H ···· F-(C_nF). This bonding is likely facilitated by an increased ionic nature of the C-F bond in fluoronanotubes in contrast with alkyl fluorides, in which, according to recent studies, the fluorine is suggested to be a poor hydrogen bond acceptor [40].

Insights into the possible fluoronanotube structures have been provided by the molecular modeling calculations and scanning tunneling microscopy (STM) imaging [41]. Based on the experimentally established C_2F stoichiometry at which fluorinated tubes can still maintain their tubular structure, two isomeric structures, resulting from fluorine addition to either (1,2) or (1,4) positions within a hexagonal ring on the graphene-like side wall of the nanotube, have been proposed. The STM imaging studies show the fluorinated regions to appear as bands around the circumference of the tube [41]. This may imply that the addition of fluorine to the side wall of the pristine SWNT should occur more favorably around the circumference of the tube. Since the calculated at various levels of theory energy difference between the 1,2- and 1,4-isomers is small [41,42], both types of fluorine addition can occur simultaneously during the fluorination process and form discrete isomeric domains on the nanotube.

Studies of thermal decomposition of C_2F fluoronanotubes in vacuo at temperatures of up to 800 °C by VTP-MS [36] indicated that the pyrolysis of fluoronanotubes is dominated by formation of volatile carbon-fluorine-containing molecules and not just the loss of elementary fluorine. Along with the observation of the extended dark bands in the STM images of the partially fluorinated SWNTs C_nF ($n > 2$) due to bare nanotube side walls, these results have led to the idea of using fluorination followed by pyrolysis as chemical "scissors" for cutting the nanotubes. Recently, Gu et al. [43,44] have implemented this idea into a process which involves the fluorination of catalyst-free purified HiPco-SWNTs to a stoichiometry of C_5F followed by pyrolysis at temperatures up to 1000 °C in an argon atmosphere that results in cutting the nanotubes at the fluorinated sites to short lengths (20-200 nm). When the pyrolysis process was monitored in situ with TGA-FTIR, the fluorine was shown to be driven from the fluoronanotube structure as CF_4 and COF_2 (due to oxygen covalently attached to the SWNT side-wall during the first steps of the purification procedure [45]). The AFM images have shown a drastically shortened lengths in comparison with the pristine SWNTs.

The preparation of hydrogenated carbon nanotubes (H-NTs) has proven to be successful by using purified carbon nanotubes and a reduction method, utilizing Li and methanol in liquid ammonia [46]. Two types of nanotubes, L-SWNTs and MWNTs, have been hydrogenated to approximately $C_{11}H$ composition determined from the yield of hydrogen evolved at elevated temperatures in the course of thermogravimetry-mass spectrometry (TG-MS) studies. Thus obtained H-NTs are thermally stable up to 400 °C. Above this temperature the evolution of hydrogen and a small amount of methane takes place, peaking at 500 °C, which provide firm evidence for covalent C-H bonds present in these hydrogenated samples. Based on these experimental data, it has been suggested that the C-H bonds in H-NTs are weaker than those in aliphatic and aromatic hydrocarbons, but can be compared with the thermal stability of C-H bonds in hydroaromatic compounds. This is supported by the density functional theory (DFT) calculations data which show that the C-H bonding energy in SWNTs hydrogenated to their maximum capacity for hydrogen storage exceeding 14 wt. % is substantially lower than in methane [47].

The idea of using the carbon nanotubes as a free radical "sponge" originated from the well-known ability of their fullerene relatives to add a variety of free radicals [5,67]. In recent studies [48,49] HiPco-SWNTs have been reacted with a series of organic peroxides, such as lauroyl, benzoyl, t-butyl and hydrogen peroxides, commonly used as radical precursors in polymerization processes. It was shown that both the pristine and fluorinated SWNTs can successfully add to their side walls a carbon-centered organic radicals, generated from acyl peroxides. The sidewall C-C bonding reactions of undecyl radicals, $C_{11}H_{23}$, generated from lauroyl peroxide, and phenyl radicals, produced from benzoyl peroxide (Scheme 1), proceeded more efficiently in case of F-SWNTs than pristine SWNTs, as indicated by a much lower reaction time (3-8 hrs vs. 5 days) [49].

Scheme 1

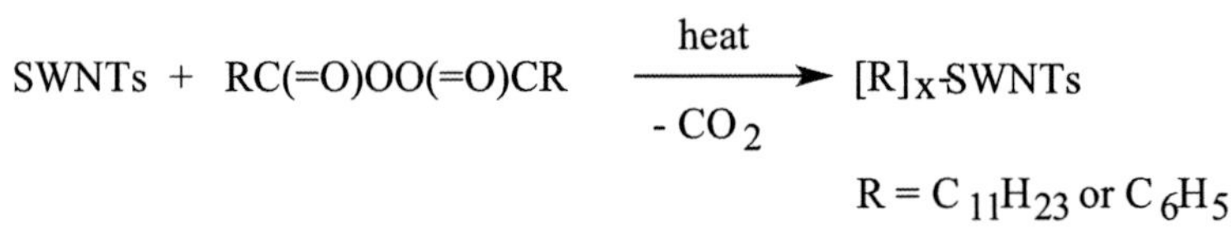

Covalent functionalization of SWNTs by this fairly simple approach using the low cost commercially available peroxides has been established by absorption UV-Vis-NIR and FTIR spectra, Raman spectroscopy, and also TG-MS and TG-FTIR analyses indicating that approximately 1 in 8 to 12 sidewall carbons carry an attached group [49].

The derivatization of SWNTs has been further developed by adding the alkyl groups terminated with a carboxylic acid moiety to the nanotubes [50]. This method, which is non-destructive to the sidewalls, utilizes the organic acyl peroxides of dicarboxylic acids, $HOOC(CH_2)_nC(O)OO(O)C(CH_2)_nCOOH$ (**1a**, n=2; **1b**, n=3), as precursors for the corresponding "functional" radicals. Succinic acid peroxide, **7a**, decomposes to form a $HOOCCH_2CH_2COO\cdot$ radical, which can subsequently lose CO_2 to yield a 2-carboxyethyl radical. Glutaric acid peroxide, **1b**, yields a 3-carboxypropyl radical via a similar route. These carboxyalkyl radicals react (Scheme 2) with the SWNTs to produce sidewall acid-functionalized SWNT-derivatives, "carboxyl" nanotubes, **2a** and **2b**, respectively.

Scheme 2

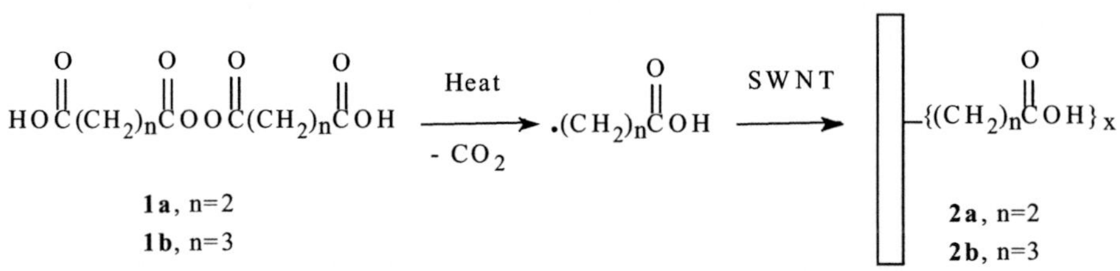

The attached carboxylic acid groups in **2a** and **2b** have been chemically characterized by formation of amide derivatives in the course of subsequent reactions with thionyl chloride and then with several terminal diamines, such as ethylenediamine, 4,4'-methylenebis (cyclohexylamine), and diethyltoluenediamine, and by instrumental characterization methods - FTIR, Raman, UV-Vis-NIR, TG-MS, TEM and solid state ^{13}C NMR [50]. The solubility data show that unlike the pristine nanotubes, the derivative **2a** is soluble in iso-propanol and water (~0.1g/L).

Recent DFT calculations of the electronic densities of states have shown that the Fermi energy of the fluoronanotubes is considerably shifted towards lower values compared to the pristine SWNTs. The conduction bands are energetically lowered as well [42]. This implies that the fluoronanotubes are better electron acceptors than the bare carbon nanotubes and therefore they should more eagerly interact with strong nucleophilic reagents as well as undergo reduction to bare SWNTs by alkali metals.

These chemical reactions are also facilitated by the fact that the C-F bonds in fluoronanotubes as in fluorinated fullerenes are weakened relative to the C-F bonds in alkyl fluorides owing to an eclipsing strain effect [51,52], and thus fluorine could be more easily displaced. The unique electronic structure and improved solubility of fluoronanotubes and the weaker C-F bond have opened new opportunities for chemical syntheses of a wide variety of sidewall functionalized nanotubes with interesting properties using fluoronanotubes as precursors.

Strong nucleophiles, such as alkyl lithium reagents, can be used to attach alkyl groups to the sidewalls of the nanotubes [7,37]. The reactions have been carried out by adding the RLi (R = methyl, n-butyl, t-butyl, n-hexyl, phenyl) reagent to the fluoronanotubes, dispersed in THF, ether, or hexane, at –40 °C followed by sonication for 10 min and stirring at room temperature overnight under argon. After completion of the workup procedure [37], the final products were dryed in a vacuum oven and then characterized by ATR-FTIR, UV-Vis-NIR, TGA and TGA-FTIR techniques.

The solubility of fluoronanotubes in alcohols prompted their functionalization reactions with alkoxides. For instance, sonication of the fluoronanotubes (~ C_2F) in methanol solution of sodium methoxide for 2 hrs resulted in the formation of methoxylated tubes with the stoichiometry of $C_{4.4}F(OCH_3)_{0.25}$ suggested from electron probe microanalysis (EPMA) [36]. The functionalizations of the fluorinated SWNTs by methoxy, ethoxy and iso-propoxy groups were also done by sonication in lithium hydroxide solutions in methanol, ethanol and iso-propanol, respectively [7]. Based on the established high reactivity of F-SWNTs towards alkoxides, a simple one-step chemical method for preparation of the SWNTs functionalized with the hydroxyl group terminated moieties ("hydroxyl nanotubes") has been developed. Such method of attaching the functional groups through the sidewall C-O bonding (Scheme 3) involved reactions of F-SWNTs with a series of alkane diols and triols pre-treated with the alkali bases MOH [53]. The series included ethylene glycol (**3a**), 1,3-propanediol (**3b**), 1,4-butanediol (**3c**), 1,2-propanediol (**3d**), 1,2-butanediol (**3e**), and glycerol (**3f**). The alcohols, diols and glycerol, were used both as solvent media and as reagents to provide a surplus of hydroxyl-terminated monoalkoxides through the reactions with alkali bases.

Scheme 3

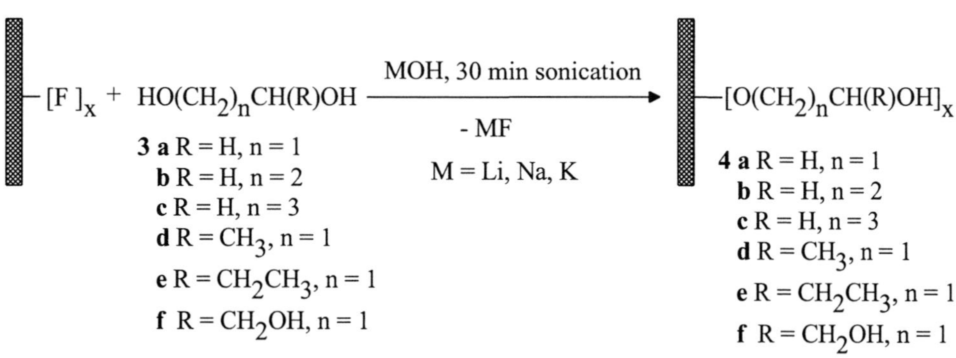

The synthesized "hydroxyl nanotubes" **4a-f** were characterized by optical spectroscopy, ATR-FTIR (Figure 1), Raman, and UV-Vis-NIR, TEM (Figure 2), AFM, and thermal degradation – TGA and VTP-MS materials characterization methods. The characterization of "hydroxyl nanotube" derivatives by Raman spectroscopy has provided essential and quick information for evaluation of the covalent sidewall modification of the nanotubes. The Raman spectra collected for all of these SWNT derivatives display a peak in the 1285-1300-cm^{-1} region related to the sp^3 states of carbon and serve as a sure proof of the disruption of the aromatic system of π-electrons on the nanotube sidewalls by the attached functional groups.

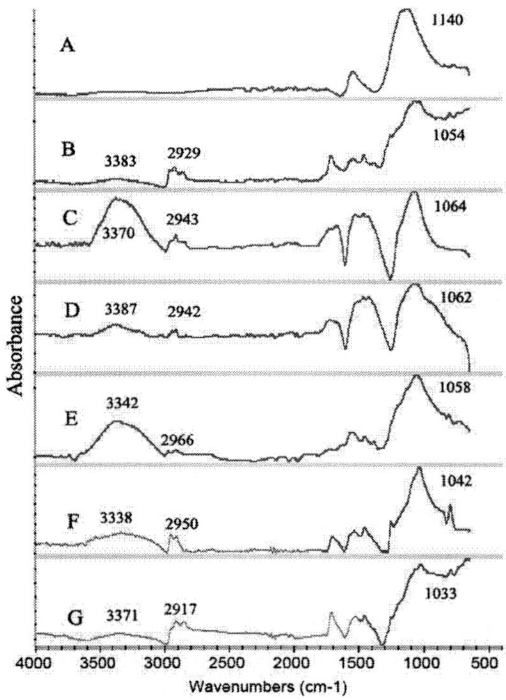

Figure 1. ATR-FTIR spectra of fluoronanotubes (A) and "hydroxyl nanotubes" (B) **4a,** (C) **4b**, (D) **4c**, (E) **4d**, (F) **4e**, (G) **4f**, showing characteristic absorption bands of functional groups attached to the sidewalls of SWNTs.

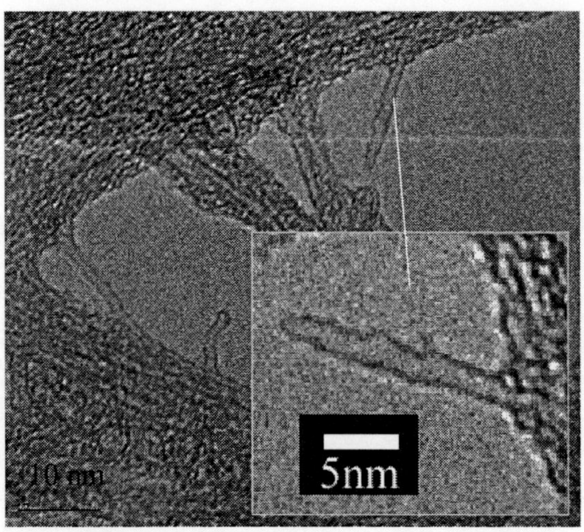

Figure 2. TEM image of the specimen of glycerol functionalized "hydroxyl nanotubes". Inset shows a zoomed-on picture of a single functionalized nanotube.

The EDX elemental analyses showed 3-5 at. % residual fluorine content in these derivatives. The degree of sidewall functionalization in **4a-f** was estimated to be in the range of 1 in 15 to 25 carbons, depending on alcohol reagent used. All prepared "hydroxyl nanotube" SWNT derivatives have shown an improved solubility in polar solvents as compared with pristine SWNTs. The solutions of glycerol-SWNT derivative **4f** in water (~ 40 mg/L) were stable for several days, while ethanol solutions with higher **4f** concentration (~80 mg/L) showed little precipitation only after several months (Figure 3). The applications of so prepared SWNT derivatives will be based on hydrogen bonding ability and chemical reactivity of terminal hydroxyl groups in the side chain.

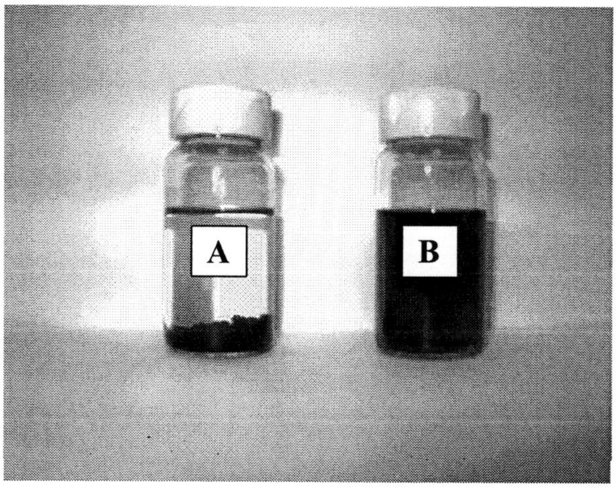

Figure 3. Photographs of the SWNT materials dispersion in ethanol: (A) pristine SWNT and (B) glycerol functionalized "hydroxyl nanotubes".

The amino group terminated SWNT derivatives, "amino nanotubes", have been prepared by interactions of fluoronanotubes with the terminal diamines, $H_2N(CH_2)_nNH_2$ (n=2,3,4,6) [54,55]. The reactions (Scheme 4) were carried out by refluxing the F-SWNTs in the corresponding diamine for 3 hrs in the presence of a catalytic amounts of pyridine (Py).

Scheme 4

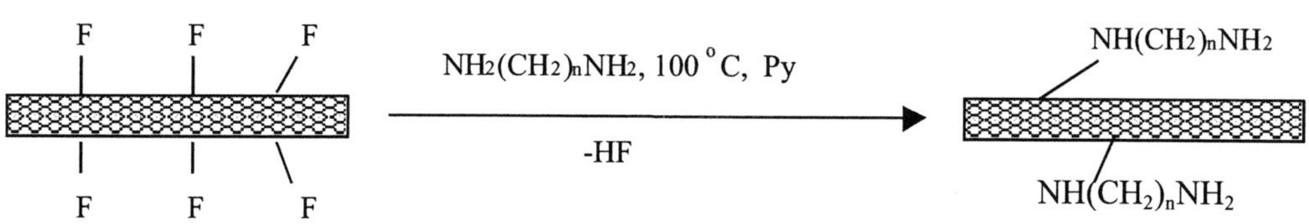

The EDX analyses of black precipitates yielded a nitrogen content within 11-16 at. %, and a very low (1-2 at. %) fluorine content, suggesting its efficient displacement by the N-alkylamino functionalities. All prepared amino-SWNT derivatives have been tested positively by Kaiser testing procedure for the free NH_2 groups. The TGA-FTIR, Raman and VTP-EIMS studies provided strong evidences for covalent functionalization, showing a major loss of the attached groups at ~325 °C [54,55].

Since diamines have shown the ability to react with fluoronanotubes by sidewall C-N bond forming interaction, the amino alcohols were also chozen to react with the fluoronanotubes in the presence of pyridine through their NH_2- rather than OH- end groups. By this way, the method of F-SWNTs sidewall C-N functionalization has been extended to the other amino group-terminated reagents. For instance, the reactions of fluoronanotubes with amino alcohols (Scheme 5), such as 2-aminoethanol **(5a)**, 3-aminopropanol **(5b),** and diethanol amine **(5c),** in the presence of pyridine as catalyst consequently yield the hydroxyl group-terminated SWNT derivatives **6a-c**, thus providing another method for synthesis of "hydroxyl nanotubes" [53].

Scheme 5

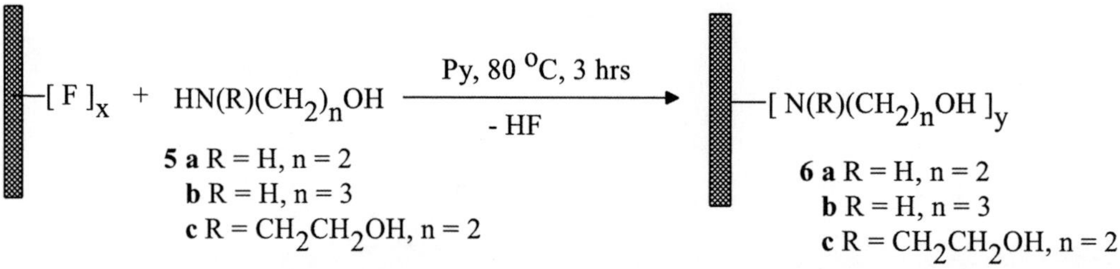

Unlike these reactions, treatment of F-SWNTs with the simplest diamine, hydrazine H_2NNH_2 resulted in their complete defluorination via the following reaction: $4C_nF$ (F-SWNT) + $N_2H_4 \rightarrow 4C_n$ (SWNT) + 4HF + N_2. [7] This reaction provides a useful tool for chemical modification of the side walls of the SWNTs, which can be applied for controlled partial defluorination of fluoronanotubes to produce SWNTs with varying fluorine contents and, presumably, different properties.

3 Carbon Nanotube-Polymer Composites

The organic functional groups, chemically attached to the nanotubes through a number of developed methods, assist effectively in unroping the nanotube bundles and improving the solubility, dispersion, processing and compatibility of carbon nanotubes when blended with the polymer matrix. The carbon nanotubes, functionalized with the amino-, hydroxy- and carboxylic acid terminal groups, are particularly important for further use as "polymerizable" nanostructured precursors, since they can be tailored for covalent integration into polymer matrices, e.g., Nylon, Kevlar (and other polyamides), epoxy based systems, polystyrene, polyethylene, polypropylene etc. through selective chemistry.

Indeed, the availability of terminal NH_2 groups in SWNTs, functionalized with ethylene diamine ("amino-SWNTs"), makes these nanotubes soluble in dilute acids and polar solvents and allows them to react, for example, with the dicarboxylic acids or their acid chlorides to form a new "nylon-SWNT" hybrid materials. To test this idea, the amino-SWNTs **7** have been reacted (Scheme 6) with the adipyl chloride [54].

Scheme 6

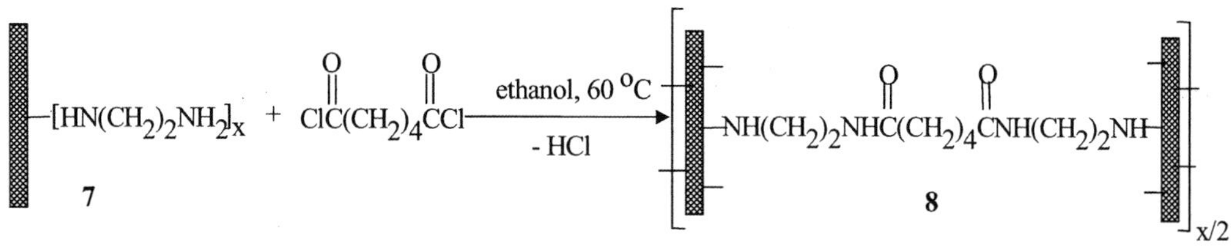

The creation of the amide -C(=O)NH- linkages in the resulting polymer **8**, formed through a cross-linking condensation reaction, was established by observation in the IR spectrum of **8** of a medium intensity bands at 3290, 1645 and 1340 cm^{-1} due to N-H, C=O and C-N stretches, typical for secondary amides.

Formation of the ladder microstructure of **8** was correlated with the observed by SEM considerable modification of the morphology of nanotube materials from thin bundles, typical of F-SWNTs (Figure 4A), to the raft-like morphology of polymer **8** (Figure 4B).

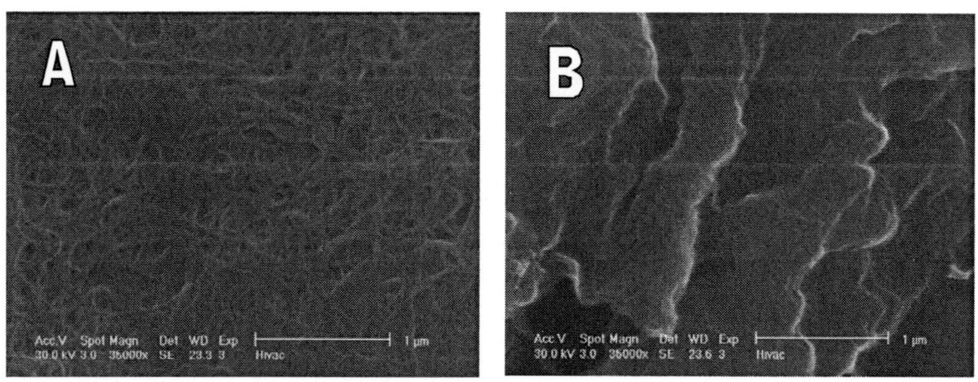

Figure 4. SEM images of F-SWNTs (A) and nylon-SWNT polymer **8** (B)

These preliminary results suggest a possibility of using the amino-functionalized SWNTs as reactive monomers for broad range of polycondensation reactions leading to new polymer materials. As an example, we have demonstrated a method for preparation of nanotube-epoxy polymer composites by using an in situ amino-functionalization route for the SWNTs [56]. Prior to loading into composites, the purified BuckyPearl (CNI) SWNTs have been first oxidized to form shortened nanotubes with the carboxyl groups located mainly on their end tips, and then fluorinated to produce F-SWNT-COOH derivatives which in comparison with pristine SWNTs (Figure 5a) showed improved dispersion both in DMF (Figure 5b) and in epoxy resin matrix – diglycidyl ether of biphenol A – EPON 862.

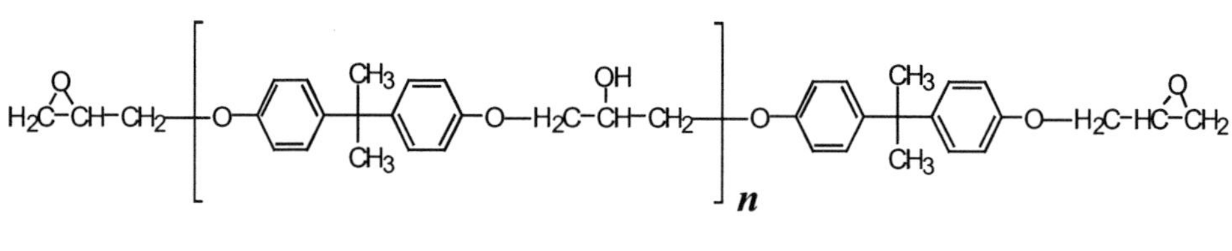

This resin was used in combination with the commercial aromatic diamine EPI-CURE W as a curing agent, capable of reacting with both epoxy rings of the resin and the C-F bonds on the side walls of SWNT derivatives. All nanotubes/epoxy composites were prepared using a 1 wt.% load of both pristine BuckyPearl SWNTs and functionalized SWNTs. For comparison, a control sample from pure epoxy was also prepared and tested.

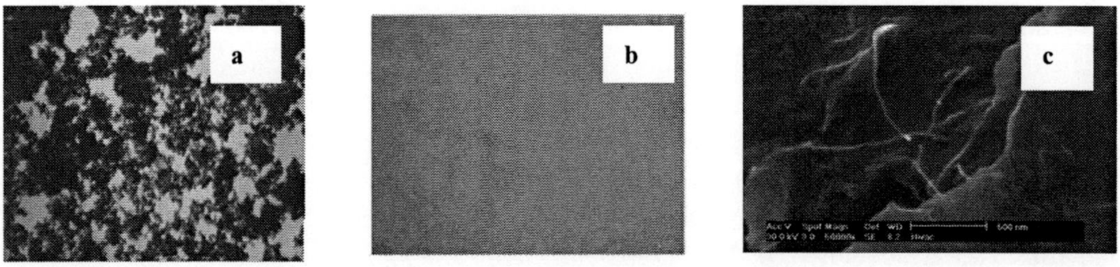

Figure 5. Optical microscopy pictures of the nanotube dispersions (2 mg/ml) in DMF: (a) pristine SWNTs, (b) F-SWNT-COOH, and (c) SEM image of fracture surface of the nanotube-epoxy composite prepared with functionalized SWNTs.

The SEM images of the composite fracture surfaces (Figure 5c) show the nanotube bundles which break rather than just pull out at the surface, suggesting that covalent bonding exists between the epoxy matrix and SWNTs. Tensile tests data (Table 1) for the epoxy composites loaded with 1 wt. % of functionalized SWNTs (F-SWNT-COOH) show a tensile strength of 95 MPa and a modulus of 2632 MPa, which means 18 and 24% improvements over the epoxy composites with pristine SWNTs and a 30% increase over that of the neat epoxy resin [56].

Table 1. Tensile properties of epoxy composites

Epoxy Composites	Young's Modulus E (Mpa)	Tensile Strength σ (MPa)
Neat Resin 862/W	1511	83.2
1% BuckyPearl SWNTs	2026	79.9
1% F-SWNT	2123	89.2
1% F-SWNT-COOH	2632	95.0

The use of the other type functionalized SWNTs with the moieties terminated by NH_2 groups for the design of fully integrated nanotube polymer composites was demonstrated by processing and fabrication of epoxy composites (EPON 862) loaded with the amide derivative of "carboxyl nanotube" **2a**, SWNT-$CH_2CH_2C(=O)NHC_6H_{10}CH_2C_6H_{10}NH_2$ **(9)** [57]. The direct impact of functionalized nanotubes **9** on the mechanical properties of the epoxy polymer has been evaluated by measurements of the tensile strength, Young's modulus and strain to failure. Epoxy composites with a small loading of **9** showed significant improvement in the mechanical properties. The average tensile strength at break increased from 83 MPa to 104 MPa, which is 25 % higher than for neat EPON 862 epoxy, and a 30 % increase as compared to the prisitine nanotube-epoxy system. The Young's modulus has shown more than 30 % improvement at just 1 wt % loading of functionalized nanotubes. For the higher loaded (4 wt % **9**) composites, up to 70 % improvement in Young's modulus was found. More attractively, the nanotube-epoxy composite samples containing 1 wt % **9** exhibited the increase of their ultimate elongation, up to 8.5 %, which is a 30 % increase as compared to the relative brittle neat epoxy polymer [57]. This presents another proof of the strong reinforcement effect in actual nanotube-filled composites considering the very high flexibility of nanotubes.

The application of "hydroxyl nanotubes" in epoxy polymer composites was also demonstrated after modification of the terminal OH groups in SWNT derivative **3b** by coupling reaction with (3-Glycidoxypropyl) trimethoxysilane [58]. The resulting silane-treated SWNTs, terminated with the glycidyl (epoxy ring) groups making the nanotubes epoxy compatible, were added (1 wt. %) to EPON 862/W epoxy resin and cured by adding a PACM aliphatic diamine curing agent. The direct effect of functionalized nanotubes on the mechanical properties of the polymer has been evaluated by DMA (Dynamic Mechanical Analysis), which gives the temperature-dependent properties of materials, such as the storage modulus. In these experiments the nanotube-epoxy composites have shown increased stiffness compared to the neat epoxy polymer (about 30% at room temperature) and the composites loaded with pristine SWNTs. The observed efficient enhancement by functionalized nanotubes resulted from both better dispersion and interfacial bonding.

Besides applications of functionalized SWNTs for reinforcing the thermoset polymer matrices (epoxies), their successful use for the same purpose in thermoplastic polymer composites was also demonstrated recently. Shofner et.al. [59,60] have studied the processing, nanotube/matrix interaction, dynamic mechanical properties, and thermal degradation of polyethylene matrix composites containing fluorinated SWNTs (F-SWNTs) in comparison with the composites containing purified SWNTs (P-SWNTs). The analysis has shown that the fluorine functional groups enhance the interaction with polyethylene, improve the dispersion of nanotubes in the matrix, and increase the dynamic mechanical properties when compared to the composites containing P-SWNTs. For instance, the 5 wt.% F-SWNT/polyethylene composite exhibited distinct improvement over the neat polyethylene with an average increase of 44% [60]. The thermal decomposition pattern of the F-SWNTs was found to be changed by the surrounding polymer matrix compared to a polymer-free condition.

Scheme 7

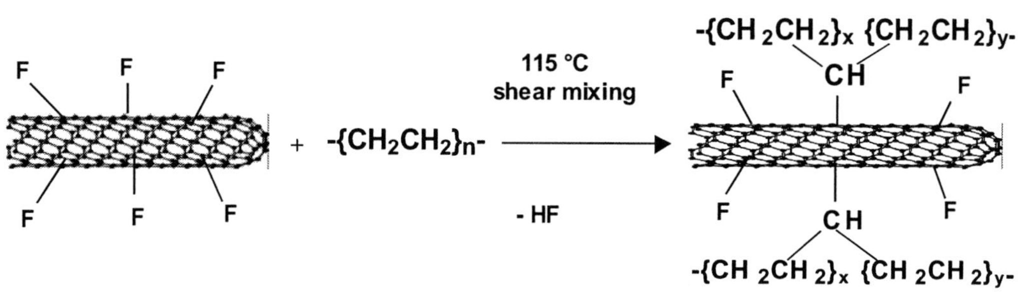

The removal of fluorine from the F-SWNTs did not lead to shortened nanotubes as in the absence of a matrix [44]. Instead, during the shear mixing melt processing the liberated fluorine groups attacked the polyethylene chain to form HF and a radical sites, creating an opportunity for grafting of the polymer to the nanotubes surface via direct covalent bonding

(Scheme 7). This result has thus demonstrated a new method for in situ design of fully integrated nanotube thermoplastic polymer composites.

4 Fluorescence Microscopy and Spectroscopy of Carbon Nanotubes

In order to visualize nanostructures one can chemically attached fluorescent molecules (fluorophores) to them. In the case of carbon nanotubes it is convenient to use the non-covalent binding fluorophores in order to preserve the nanotube pristine structure (without disturbing the sp^2 carbon-carbon bond). Pursuing the goal of depositing the isolated SWNTs onto a glass, a spin-coating technique was used with aproximately 75 μl of a sample containing SWNTs of 200–300 nm in length. In this way, one could obtain surface densities of about 10 - 20 nanotubos per 100 μm^2, which means that nanotubes will most likely not be in the contact with each other but rather isolated.

The nanotubes were deposited onto a glass substrate, using a fluorophore $DiOC_6$ solution of 1.5 μl (5mg/ml in 100% of methanol) for 99 μl of de-ionized H_2O. The non-covalent bonding of $DiOC_6$ fluorophore to carbon nanotube is likely to occur through hydrophobic interaction since the sp^2 graphene plane is hydrophobic.

One Zeiss Axiovert 100TV microscope with a CCD Hamamatsu Orca-II camera was used for observation of the sample. The images were reconstructed through phase contrast. The excitation wavelength was supplied by a cw laser of large spectrum with a 530-560 nm filter, while the emission wavelength was filtered to 590-650 nm window. Using an excitation laser of 628 nm one can assure a spectral isolation from all Raman signals, which occur between 628 and 770 nm, and from fluorescence signals with wavelengths above 850 nm.

The AFM monitoring assures that carbon nanotubes are kept isolated from each other, which is absolutely crucial, as the strong van der Waals interactions cause the clustering of nanotubes, precluding the observation of a unique exponential decay in time. The spectroscopic analysis of nanotube clusters [61,62] reveal an extremely fast de-excitation (~1ps) atributed to energy transfer among the nanotubes.

The spectroscopy of isolated nanotubes allows a detailed analysis of the optical and electronic properties, as it does not make use of sample averaging among several nanotubes. The emission spectrum displays a symmetrical band without noise and is well described by a Lorentzian distribution with a broadened bandwidth at room temperature, of aproximately 20 meV.

Electron emission was not yet observed and a possible explanation for this is the existence of a competing channel with the fluorescence decay, which is likely to be originated by nanotube defects that are able to trap electrons.

Wang et al., [61] observed a low quantum efficiency in the nanotube fluorescence (< 10^3), and a fluorescence decay time for a single electron in isolated SWNT's of about 10 ps; from these results they could deduce that $\tau_{rad} = 110$ ns.

Actually, assuming a constant decay rate for all the nanotubes, $(\tau_{rad})^{-1}$, and that at t = 0 s, $\phi(t) = (\tau_{rad})^{-1} N_0 s(t)$, where s(t) represents a fraction of excited nanotubes which remain in the emission state until the instant t, then such rate can be determined from the fluorescence decay curve, by integration of the fluorescence rate between t=0 and $t \to \infty$:

$$(\tau_{rad})^{-1} = \eta/\langle\tau_{exc}\rangle \quad (1)$$

where $\eta = (1/N_0)\int \phi(t)dt$ is the total fluorescence and $\langle\tau_{exc}\rangle / \int s(t)dt$ is the nanotube excited state lifetime. Experimentally they [61,62] obtained $\eta = 1.7\times10^{-4}$ and $\langle\tau_{exc}\rangle = 18$ps, which leads to $\tau_{rad} = 110$ ns for the SWNT lifetime.

5 Conclusions

The recent extensive studies show that carbon nanotubes can be surface functionalized by using a number of reagents. The carbon nanotube derivatives, covalently functionalized at the open ends or at the side walls, show improved solubility in common solvents, and electrical, optical and spectroscopic properties different from pristine nanotubes. For instance, the fluoro, hydroxyl, amino and carboxyl nanotubes form stable suspension solutions in polar solvents, such as iso-propanol, ethanol, dimethylformamide, o-dichlorobenzene and even water (in cases of OH and COOH derivatives) which facilitate their improved processing in copolymer and ceramics nanofabrication and provide for compatibility with biomaterials.

The chemistry of the OH, COOH, and other functional groups that were attached to the nanotubes, is so abundant that after subsequent modification the nanotube derivatives can be a versatile precursors in the fabrication of nanotube-reinforced co-polymers and ceramics as well as biomaterials. Thus, the developed functionalization chemistry already enabled the successful incorporation of SWNTs into epoxy and polyethylene polymer matrices and demonstration of substantial mechanical reinforcement of the resulting hybrid materials. The "chemical scissor" method for chemical cutting of the single-wall carbon nanotubes to 20-200 nanometer lengths, utilizing partial fluorination and pyrolysis of SWNTs, will open additional opportunities for the functionalization and studies of the length related properties of the cut nanotubes. Further development of fluorescence microscopy and spectroscopy methods for probing the carbon nanotubes will enhance the fundamental research in this field.

As a result, various applications of chemically modified carbon nanotubes, are expected to emerge from the continuing research in the interdisciplinary field of nanotechnology.

Acknowledgement

The authors are pleased to acknowledge the NASA-URETI program, ONR, Carbon Nanotechnologies, Inc., and the FCT/MCTES-Portugal POCTI/FIS/2000/43627 project for their support of research on nanoscience and nanotechnology. The following students and postdoctoral fellows are greatly appreciated for performing different parts of the experimental work: J. Stevens, V. Kiny, L. Zhang, A. Huang, P. Reverdy, and Drs. Z. Gu, H. Peng, J. Zhu, M. Shofner.

References

[1] S. Iijima, *Nature*, **354**, 56 (1991).
[2] S. Iijima, T. Ichihashi, *Nature*, **363**, 603 (1993).

[3] D. S. Bethune, C. H. Kiang, M. S. deVries, G. Gorman, R. Savoy, J. Vazquez, R. Bayers, *Nature*, **363**, 605 (1993).

[4] M. Endo, K. Takeuchi, S. Igarashi, K. Kobori, M. Shiraishi, H. W. Kroto, *Phys. Chem. Solids*, **54**, 1841 (1993).

[5] M. S. Dresselhaus, G. Dresselhaus, P. C. Eklund, *Science of Fullerenes and Carbon Nanotubes*, Academic Press, San Diego, Vol. 1 (1996).

[6] V. N. Khabashesku, J. L. Margrave, *Chemistry of Carbon Nanotubes in Encyclopedia of Nanoscience and Nanotechnology*, Ed. H. S. Nalwa, American Scientific Publishers, Vol. 1, 849 (2004).

[7] V. N. Khabashesku, W. E. Billups, J. L. Margrave Acc. *Chem. Res.* **35**, 1087 (2002).

[8] V. N. Khabashesku, J. L. Margrave, E. V. Barrera, Diamond Relat. *Mater.* **14**, 859 (2005).

[9] E. V. Barrera, *JOM*, **52**, 38 (2000).

[10] D. Pantarotto, C. D. Partodos, R. Graff, J. Hoebeke, J.-P. Briand, M. Prato, *A. Bianco J. Am. Chem. Soc.* **125**, 6160 (2003).

[11] V. N. Khabashesku, Functionalization of Carbon Nanomaterials for Bio-Medical Applications, in "Extending the life span" K. Sames, S. Sethe, A. Stolzing (eds.): LIT Verlag Münster (Hamburg, Berlin, London, c/o New York), 147 (2005).

[12] B. Vigolo, A. Penicaud, C. Coulon, C. Sauder, R. Pailler, C. Journet, P. Bernier, P. Poulin, *Science*, **290**, 1331 (2000).

[13] R. J. Chen, Y. Zhang, D. Wang, H. Dai, *J. Am. Chem. Soc.* **123**, 3838 (2001).

[14] Z. Jin, L. Huang, S. H. Goh, G. Xu, W. Ji, *Chem. Phys. Lett.* **332**, 461 (2000).

[15] M. J. O'Connell, P. J. Boul, L. M. Ericson, C. B. Huffman, Y. Huang, E. H. Haroz, K. D. Ausman, R. E. Smalley, *Chem. Phys. Lett.* **342**, 265 (2001).

[16] A. Star, J. F. Stoddart, D. Steuerman, M. Diehl, A. Boukai, E. W. Wong, X. Yang, S.-W. Chung, H. Choi, J. R. Heath, *Angew. Chem. Int. Ed.* **40**, 1721 (2001).

[17] A. B. Dalton, C. Stephan, J. N. Coleman, B. McCarthy, P. M. Ajayan, S. Lefrant, P. Bernier, W. J. Blau, H. J. Byrne, *J. Phys. Chem. B*, **104**, 10012 (2000).

[18] J. L. Bahr, J. M. Tour, *J. Mater. Chem.* **12**, 1952 (2002).

[19] J. Chen, M. A. Hamon, H. Hu, Y. Chen, A. M. Rao, P. C. Eklund, R. C. Haddon,. *Science*, **282**, 95 (1998).

[20] S. C. Tsang, P. J. F. Harris, M. L. H. Green, *Nature*, **362**, 520 (1993).

[21] P. M. Ajayan, T. M. Ebbesen, T. Ichihashi, S. Iijima, K. Tanigaki, H. Hiura, *Nature*, **362**, 522 (1993).

[22] J. Liu, A. G. Rinzler, H. Dai, J. H. Hafner, R. K. Bradley, P. J. Boul, A. Lu, T. Iverson, K. Shelimov, C. B. Huffman, F. Macias-Rodriguez, Y.-S. Shon, T. R. Lee, D. T. Colbert, R. R. Smalley. *Science*, **280**, 1253 (1998).

[23] A. G. Rinzler, J. Liu, H. Dai, P. Nikolaev, C. B.Huffman, F. J. Rodriguez-Macias, P. J. Boul, A. H. Lu, D. Heymann, D. T. Colbert, R. S. Lee, J. E. Fischer, A. M. Rao, P. C. Eklund, R. E Smalley, *Appl. Phys. A*, **67**, 29 (1998).

[24] I. W. Chiang, B. E. Brinson, R. E. Smalley, J. L. Margrave, R. H. Hauge, *J. Phys. Chem. B*, **105**, 1157 (2001).

[25] J. Chen, A. M. Rao, S. Lyuksyutov, M. E. Itkis, M. A. Hamon, H. Hu, R. W. Cohn, P. C. Eklund, D. T. Colbert, R. E. Smalley, R. C. Haddon, *J. Phys. Chem. B*, **105**, 2525 (2001).

[26] C. G. Bac, S. Latl, V. Jourdain, R. Aznar, F. Rachdi, P. Bernier, M. Holzinger, A. Hirsch, C. Mathis, P. Petit, A. Rubio, *In Proc. Int. Winterschool on Electronic Properties of Molecular Nanostructures*, Kirchberg, 2001, Eds. H. Kuzmany et al., AIP Conference Proceedings, Melville, NY, 2001.

[27] W. Zhou, Y. H. Ooi, R. Russo, P. Papanek, D. E. Luzzi, J. E. Fischer, M. J. Bronikowski, P. A. Willis, R. E. Smalley, *Chem. Phys. Lett.*, **350**, 6 (2001).

[28] A. M. Rao, J. Chen, E. Richter, U. Schlecht, P. C. Eklund, R. C. Haddon, U. D. Venkateswaran, Y. K. Kwon, D. Tomanek, *Phys. Rev. Lett.*, **86**(2001).

[29] D. V. Mawhinney, V. Naumenko, A. Kuznetsova, J. T. Yates, J. Liu, R. E. Smalley, *J. Am. Chem. Soc.*, **122**, 2382 (2000).

[30] Z. Liu, Z. Shen, T. Zhu, S. Hou, L. Ying, *Langmuir*, **16**, 3569 (2000).

[31] A. Thess, R. Lee, P. Nikolaev, H. Dai, P. Petit, J. Robert, C. Xu, Y. H. Lee, S. G. Kim, A. G. Rinzler, D. T. Colbert, G. E. Scusseria, D. Tomanek, J. E. Fischer, R. E. Smalley, *Science*, **273**, 483(1996).

[32] P. Nikolaev, M. J. Bronikowski, R. K. Bradley, F. Rohmund, D. T. Colbert, K. A. Smith, R. E. Smalley, *Chem. Phys. Lett.* **313**, 91(1999).

[33] M. J. Bronikowski, P. A. Willis, D. T. Colbert, K. A. Smith, R. E. Smalley, *J. Vac. Sci. Technol. A*, **19**, 1800 (2001).

[34] R. J. Lagow, R. B. Badachape, J. L.Wood, J. L. Margrave, *J. Chem. Soc. Dalton Trans.* **12**, 1268(1974).

[35] E. T. Mickelson, C. B. Huffman, A. G. Rinzler, R. E. Smalley, R. H. Hauge and J. L. Margrave, *Chem. Phys. Lett.* **296**, 188 (1998).

[36] E. T. Mickelson, *Novel Chemistry of Elemental Carbon: Graphite, Fullerenes and Nanotubes*. Ph. D. Thesis, Rice University, Houston, TX, 1999.

[37] I. W. Chiang, *Science of Single-Wall Carbon Nanotubes: Purification, Characterization and Chemistry*. Ph. D. Thesis, Rice University, Houston, TX, 2001.

[38] T. Hayashi, M. Terrones, C. Scheu, Y. A. Kim, M. Ruhle, T. Nakajima, M. Endo, *NanoLett.* **2**, 491 (2002).

[39] E. T. Mickelson, I. W. Chiang, J. L. Zimmerman, P. J. Boul, J. Lozano, J. Liu, R. E. Smalley, R. H. Hauge, J. L. Margrave, *J. Phys. Chem. B*, **103**, 4318 (1999).

[40] J. D. Dunitz, R. Taylor, *Chem. Eur. J.*, **3**, 89 (1997).

[41] K. F. Kelly, I. W. Chiang, E. T. Mickelson, R. H. Hauge, J. L. Margrave, X.Wang, G. E. Scusseria, C. Radloff, N. J. Halas, *Chem. Phys. Lett.*, **313**, 445 (1999).

[42] K. N. Kudun, H. F. Bettinger, G. E. Scuseria, *Phys. Rev. B*, **63**, 45413 (2001).

[43] Z. Gu, H. Peng, J. L. Zimmerman, I. W. Chiang, V. N. Khabashesku, R. H. Hauge, J. L. Margrave, 223rd ACS National Meeting, April 7-11, 2002, Orlando, FL. *Abstracts and Papers. Division of Fluorine Chemistry*. Abstr. p. 12.

[44] Z. Gu, H. Peng, R. H. Hauge, R. E. Smalley, *J. L. Margrave*, **2**, 1009 (2002).

[45] I. W. Chiang, B. E. Brinson, A. Y. Huang, P. A. Willis, M. J. Bronikowski, J. L. Margrave, R. E. Smalley, R. H. Hauge, *J. Phys. Chem. B*, **105**, 8297 (2001).

[46] S. Pekker, J.-P. Salvelat, E. Jakab, J.-M. Bonard, L. Forro, *J. Phys. Chem. B*, **105**, 7938 (2001).

[47] S. M. Lee, Y. H. Lee, *Appl. Phys. Lett.*, **76**, 2877 (2000).

[48] P. Reverdy, H. Peng, V. N. Khabashesku, J. L Margrave, *Rice Quantum Institute Fifteenth Annual Summer Research Colloquium*. Aug. 17, 2001, Abstr. p.19.

[49] H. Peng, P. Reverdy, V. N. Khabashesku, J. L. Margrave, *Chem. Commun.* **362** (2003).

[50] H. Peng, L. B. Alemany, J. L. Margrave, V. N. Khabashesku, *J. Am. Chem. Soc.* **125**, 15174 (2003).

[51] R. Taylor, *Russ. Chem. Bull.*, **47**, 823 (1998).

[52] H. F. Bettinger, K. N. Kudun, G. E. Scuseria, *J. Am. Chem. Soc.*, 123, 12849 (2001).

[53] L. Zhang, V. U. Kiny, H. Peng, J. Zhu, R. F. M. Lobo, J. L. Margrave, V. N. Khabashesku, *Chem. Mater.* **16**, 2055 (2004).

[54] J. L. Stevens, A. Y. Huang, H. Peng, I. W. Chiang, V. N. Khabashesku, J. L. Margrave, *Nano Lett.* **3**. 331 (2003).

[55] J. L. Stevens, V. Kiny, A. Y. Huang, I. W. Chiang, G. A. Derrien, V. N. Khabashesku, J. L. Margrave, Proc. NanoTech Vol. 3, 169 (2003).

[56] J. Zhu, J. Kim, H. Peng, J. L.Margrave, V. N. Khabashesku, E. V. Barrera, *NanoLett.* **3**, 1107 (2003).

[57] J. Zhu, H. Peng, F. Rodriguez-Macias, J. L. Margrave, V. N. Khabashesku, A. Imam, K.Lozano, E. V. Barrera, *Adv. Funct. Mater*, **14** (2004) 643.

[58] J. Zhu, V. N. Khabashesku, M. A. Imam, K. Lozano, R. Crane, E. V. Barrera, Mater. Science Forum. *TransTech Publ.*, Schwitzerland, Vols. 475-479, 1059 (2005).

[59] M. L. Shofner, H. Peng, Z. Gu, V. N. Khabashesku, J. L. Margrave, E. V. Barrera. In book “Mechanical Properties of Nanostructured Materials and Nanocomposites” *MRS Symposium Proceedings*, *Paper Number:* **Q10.9**-50936, Vol. 791, (2004).

[60] M. L. Shofner, V. N. Khabashesku, E. V. Barrera, *Chem. Mater.*, **18** (2006), 906.

[61] F. Wang, G. Dukovic, L. E. Brus, T. F. Hein, *Phys. Rev. Lett.*, **92** (2004) 177401.

[62] A. Hartschuh, H. N. Pedrosa, L. Novotny, T. D. Krauss, *Science* **301** (2003) 1354.

In: Focus on Nanotube Research
Editor: Delores A. Martin, pp. 19-36

ISBN 1-59454-851-X

Chapter 2

ON THE PAIRING MECHANISM IN DOPED ARMCHAIR CARBON NANOTUBES

Enrico Perfetto and Michele Cini
Istituto Nazionale per la Fisica della Materia, Dipartimento di Fisica,
Universita' di Rome Tor Vergata, Via della Ricerca Scientifica 1,
00133 Roma Italy and INFN - Laboratori Nazionali di Frascati,
Via E. Fermi 40, 00044 Frascati, Italy
Gianluca Stefanucci
Department of Solid State Theory, Institute of Physics,
Lund University, Sölvegatan 14 A, 22362 Lund Sweden

Abstract

We review an electronic pairing mechanism driven by correlation, i.e. by the on-site repulsion; it works in a Hubbard model with the symmetry of the wrapped honeycomb lattice, suitable for single-wall (N, N) carbon nanotubes. There is an important analogy with the one-band and three-bands Hubbard model for Cuprates: the Hubbard Hamiltonian for nanotubes also admits two-body singlet eigenstates with vanishing on-site repulsion, which we call $W = 0$ pairs. By means of a non-perturbative canonical transformation we calculate the effective interaction between the electrons of a $W = 0$ pair added to the interacting ground state. The interesting possibility of bound states is borne out by our approach provided the nanotube is doped away from half filling. Then, the *dressed* $W = 0$ pair in the interacting many-body system is a bound state, and this occurs with quite reasonable parameter values. The pairing mechanism uses electronic degeneracy (or quasi-degeneracy) to produce attraction from repulsion. Pairing in Hubbard models has been reported by other authors as well, but the present approach has certain specific advantages. For example, it shows how pairing is favored by a large noncommutative symmetry Group (the Space Group, possibly supplemented by extra operations). This finding strongly suggests that the recently discovered superconductivity in carbon nanotubes could be at least partially due to an electronic mechanism; the physics involved could have some features in common with the case of High T_c compounds. From the effective interaction in the nanotubes we compute pair binding energies; we use exact diagonalization results for the (1,1) nanotube to further confirm the analytic theory, then we consider the general trend. There is an optimal doping when the number of electrons per C atom is in the range 1.2÷1.3. For (N, N) nanotubes of length L, the binding energy of the pair depends strongly on the filling and decreases towards a limiting value of the order of $0.1 \div 1$ meV as $L \to \infty$. We also present preliminary results on the inclusion of phonons in the present theory of pairing in the Hubbard model. Due to the need to include several oscillator branches the problem is hard and we resort to small cluster calculations, to start forming broad criteria. The results indicate that a strong Jahn-Teller effect induced by vector phonons may destroy the symmetry; the correlation-induced pairing, which is driven by the symmetry, is broken in such cases. However we also show examples where electronic and phonon-induced pairing are synergic, both at weak coupling and in the polaronic regime, and the distortion is dynamically avoided.

1 Introduction

A single-wall carbon nanotube (SWNT)[1] is a graphite sheet wrapped onto a cylinder. The carbon atoms are arranged on the sites of a honeycomb lattice and the chirality of the wrapping determines the electronic properties of the tube. In particular the two main chiralities give rise to metallic ("armchair") and semiconducting ("zig-zag") SWNT's. Metallic "armchair" tubes have four linear branches crossing the two Fermi points and the higher subbands, characterized by a nonvanishing transverse momentum, lie about $0.1 \sim 0.2$eV above. Therefore, in the very low-energy limit, they can be considered as strictly one dimensional systems. As a consequence there is general agreement in addressing their electronic properties to the Luttinger liquid class of universality[2][3].

However such a low-energy scenario may be significantly changed if the nanotube is driven away from half filling, causing the instability of the Luttinger liquid fixed point. Indeed there has been growing evidence of superconducting fluctuations in SWNT's placed between superconducting contacts[4][5][6] up to the transition temperature of $\simeq 0.5$ K[7].

It was also suggested that by a suitable injection of charge carriers, the density of states at the Fermi level may be increased, with a enhancement of the superconducting correlations. In a strongly doped regime the Fermi level is pushed away from the undoped position; as a consequence subbands with nonvanishing transverse momentum are involved in the low-energy scattering processes and a simple one dimensional description cannot be applied. In other terms, the transverse direction must be taken into account. The shift of the Fermi level can be achieved by two basic methods, i.e. by tuning an additional gate voltage or by intercalating alkali impurities. The electrochemical doping of ropes of carbon nanotubes has been reported in recent experiments[8], where Litium was intercalated in the interstitial sites between the tubes arranged in a triangular lattice, up to a concentration corresponding to $Li_{1.7}C_6$. First priciple calculations performed on this system showed that an almost complete charge transfer occurs between Li atoms and C atoms[9]. Furthermore the resulting deformation of the nanotubes is predicted to be very small and as a consequence the tranferred electrons mainly occupy the original bands of the carbon nanotubes. Finally the effect of doping consists on shifting the Fermi level about 1eV above the initial position and the electronic properties of the individual tubes result deeply modified.

Here we review an electronic pairing mechanism suitable for highly doped SWNT's [10][11]. We describe the tube with a Hubbard model defined on the wrapped honeycomb lattice, disregarding the long-range character of the Coulomb interaction. The long-range part of the interaction in nanotubes has been shown recently [12] to have a marked influence

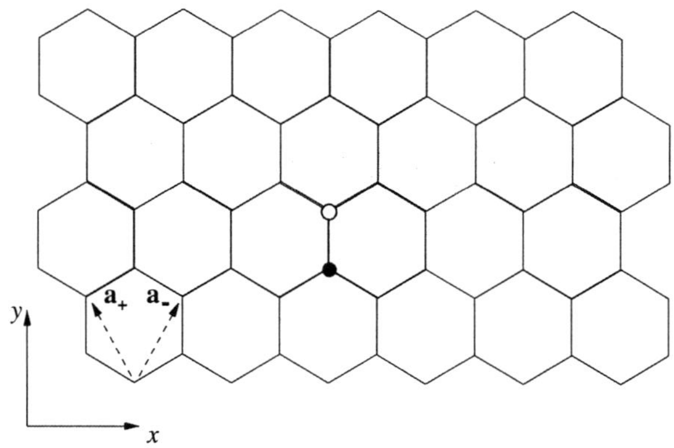

Figure 1: Illustration of the honeycomb lattice. The sites a (empty dots ∘) and b (full dots •) constitute the basis of each unit cell. The (N, N) nanotube is obtained by identifying sites separated by $N\mathbf{a}_+ + N\mathbf{a}_-$; the axis of the tube is along the x direction.

on the Coulomb blockade pattern, but can be safely disregarded here due to the small size of the pairs. The key point relies in identifying the Cooper pairs with the so called $W = 0$ pairs. A $W = 0$ pair is defined as an *exact* two-body singlet eigenstate of the Hubbard Hamiltonian without double occupancy. This means that a $W = 0$ pair is at the same time an eigenstate of the kinetic energy and of the Hubbard repulsion with vanishing eigenvalue for the latter.

$W = 0$ pairs arise in the presence of high enough symmetry, and provide a non-conventional mechanism of pairing from repulsive interactions in widely different materials. In recent years, we proposed a symmetry-driven configuration interaction pairing mechanism [13][14][15] in the *two-dimensional* (one-band and three-bands) repulsive Hubbard model for the Cuprates, showing that the $W = 0$ pairs, once dressed by the interaction, become bound states. Here, we wish to extend the approach in order to deal with the SWNT. We present analytic expressions for the effective interaction and obtain the binding energy for (N, N) armchair nanotubes [see Fig.(1)]; in the case $N = 1$ we verify the analytic results by exact diagonalization and get high-precision agreement. Some experimental trends are correctly reproduced. In particular, a reasonable energy scale for the superconducting gap arises from the ~ 1eV input parameters and a stronger pairing is correctly predicted for smaller radii; this agrees with the measurements in SWNT by Tang *et al.*[5].

We get effective attraction and pairing for doped nanotubes, and starting from the undoped system we find that the pair binding energy grows as the number of electrons per C atom increases until an *optimum* is reached.

2 $W = 0$ Pairs in the Hubbard Model

The Hubbard Hamiltonian H on the wrapped honeycomb lattice for the valence bands of (N, N) SWNT's reads:

$$H = H_0 + W = t \sum_{\langle \mathbf{r},\mathbf{r}' \rangle} \sum_{\sigma} \left(c^{\dagger}_{\mathbf{r},\sigma} c_{\mathbf{r}',\sigma} + h.c. \right) + U \sum_{\mathbf{r}} \hat{n}_{\mathbf{r},\uparrow} \hat{n}_{\mathbf{r},\downarrow}, \tag{1}$$

where $\mathbf{r}$ denotes the honeycomb site, the sum runs over the pairs $\langle \mathbf{r}, \mathbf{r}' \rangle$ of nearest neighbour carbon atoms, t is the hopping parameter and U is the on-site Hubbard repulsion. The one-body eigenvalues $\varepsilon^{\pm}(\mathbf{k})$, (- for the bonding and + for the antibonding bands) are readily obtained[10], and the Fermi line has C_{2v} symmetry for the nanotubes (C_{6v} for the graphite sheet). Here, we assume that the Fermi level ε_F lies in the + band as for alkali-doped compounds.

The particles forming a $W = 0$ pair have no direct interaction and are the most natural candidates to achieve bound states in purely repulsive Hubbard models. We note that such states are also building bricks of the ground state of Hubbard and related models at half filling[16][17][18]. Recently we obtained[15][19] a general criterion to get *all* the $W = 0$ pairs for Hubbard models defined on any graph. In the case of triplet pairs the absence of the on-site interaction in the Hubbard model is a trivial consequence of the Pauli principle; the point that we wish to make here is that the extension to singlet eigenstates of the Hamiltonian can only be a consequence of symmetry and is possible only when enough symmetry is present. When such solutions exist, we can fully classify them and predict their properties by the use of Group theory. The square or hexagonal point symmetry is not generally enough. We must speak in terms of the Optimal Group $\mathcal{G}_0$ of the Hamiltonian, that we define as a symmetry Group that has enough operations to justify the degeneracy of the single particle energy levels. We have relegated all the technicalities in the next subsection, which may be skipped by those readers who are unfamiliar with Group Theory or are satisfied with an intuitive understanding of the argument; for those readers, the essence of the next Subsection is that $W = 0$ pairs belong to well defined symmetry types, that are completely different from those where single particle solutions of the same Hamiltonians exist. The theorem also tells us that the *bigger* is $\mathcal{G}_0$, the *larger* is the number of $W = 0$ pairs. At the same time, a big "non-interacting" symmetry Group $\mathcal{G}_0$ grants large level degeneracies. As we have seen, this fact allows the existence of $W = 0$ pairs formed by degenerate orbitals.

2.1 A Useful Theorem

Let $\mathcal{G}_0$ be the the symmetry Group of the non-interacting Hubbard Hamiltonian $K = \sum_{\langle ij \rangle, \sigma} t_{ij} c^{\dagger}_{j\sigma} c_{i\sigma}$ defined on an arbitrary graph Λ. By definition no degeneracy between one-body eigenstates is *accidental*, hence $\mathcal{G}_0$ must contain enough operations to justify all such degeneracies. Let us label each one-body eigenstate of $H = K + W$, $W = \sum_{i \in \Lambda} U_i n_{i\uparrow} n_{i\downarrow}$, with an irreducible representation (irrep) of $\mathcal{G}_0$.

Definition. An irrep η is **represented** in the one-body spectrum of H if at least one of the one-body levels belongs to η.

Let $\mathcal{E}$ be the set of the irreps of $\mathcal{G}_0$ which are represented in the one-body spectrum of H. Let $|\psi\rangle$ be a two-body eigenstate of the non-interacting Hamiltonian with spin $S^z = 0$ and $P^{(\eta)}$ the projection operator on the irrep η.

We wish to prove the

<u>$W = 0$ *Theorem.*</u> - Any nonvanishing projection of $|\psi\rangle$ on an irrep <u>not</u> contained in $\mathcal{E}$, is an eigenstate of H with no double occupancy. The singlet component of this state is a $W = 0$ pair. Conversely, any pair belonging to an irrep represented in the spectrum must have non-vanishing W expectation value:

$$\eta \notin \mathcal{E} \Leftrightarrow WP^{(\eta)}|\psi\rangle = 0. \tag{2}$$

Remark: Suppose we perform a gauge change in H such that $\mathcal{G}_0$ is preserved; clearly, a $W = 0$ pair goes to another $W = 0$ pair. Thus, the theorem implies a distinction between symmetry types which is gauge-independent.

Proof: Let us consider a two-body state of opposite spins belonging to the irrep η of $\mathcal{G}_0$:

$$|\psi^{(\eta)}\rangle = \sum_{ij\in\Lambda} \psi^{(\eta)}(i,j) c^\dagger_{i\uparrow} c^\dagger_{j\downarrow}|0\rangle.$$

Then we have

$$n_{i\uparrow} n_{i\downarrow}|\psi^{(\eta)}\rangle = \psi^{(\eta)}(i,i) c^\dagger_{i\uparrow} c^\dagger_{i\downarrow}|0\rangle \equiv \psi^{(\eta)}(i,i)|i\uparrow, i\downarrow\rangle.$$

We define $P^{(\eta)}$ as the projection operator on the irrep η. By definition

$$P^{(\eta)} \sum_{i\in\Lambda} \psi^{(\eta)}(i,i)|i\uparrow, i\downarrow\rangle = \sum_{i\in\Lambda} \psi^{(\eta)}(i,i)|i\uparrow, i\downarrow\rangle,$$

and therefore $\psi^{(\eta)}(i,i) = 0,\ \forall i \in \Lambda$, provided $P^{(\eta)}|i\uparrow, i\downarrow\rangle = 0,\ \forall i \in \Lambda$. It is worth noting that this condition is satisfied if and only if $P^{(\eta)}|i\sigma\rangle = 0,\ \forall i \in \Lambda$, where $|i\sigma\rangle = c^\dagger_{i\sigma}|0\rangle$. Now it is always possible to write $|i\sigma\rangle$ in the form $|i\sigma\rangle = \sum_{\eta\in\mathcal{E}} c^{(\eta)}(i)|\eta_\sigma\rangle$ where $|\eta_\sigma\rangle$ is the one-body eigenstate of H with spin σ belonging to the irrep η. Hence, if $\eta' \notin \mathcal{E}$, $P^{(\eta')}|i\sigma\rangle$ vanishes and so does $P^{(\eta')}|i\uparrow, i\downarrow\rangle$. Therefore, if $|\psi^{(\eta)}\rangle$ is a two-hole singlet eigenstate of the kinetic term and $\eta \notin \mathcal{E}$, then it is also an eigenstate of W with vanishing eigenvalue, that is a $W = 0$ pair.

Q.E.D.

As anticipated above, the theorem also tells us that the *bigger* is $\mathcal{G}_0$, the *larger* is the number of $W = 0$ pairs. Indeed, for a given system, the number of one-body eigenvectors is fixed, while the number and the dimension of the irreps grow with the order of the group. Therefore, also the number of irreps not represented in $\mathcal{E}$ grows, meaning more $W = 0$ pairs.

Moreover a partial use of this theorem is possible if we know only a subgroup $\mathbf{G_0} \subset \mathcal{G}_0$. In this case the "$\Rightarrow$" implication still holds for the irreps of $\mathbf{G_0}$.

2.2 Symmetry of $W = 0$ Pairs for Carbon Natotubes

For the (N, N) SWNT, we used the Space Group as $\mathbf{G_0}$. We considered the case $N > 1$ and we focused on the subspace of vanishing quasi-momentum. We found that the only $W = 0$ pairs belong to the pseudoscalar irrep A_2. Let (a, b) denote the basis of the Bravais lattice [see Fig.(1)] and $u(\mathbf{k}, \zeta)$ the periodic part of the Bloch function of quasi-momentum $\mathbf{k}$, with $\zeta = a, b$. The pair wavefunction reads[10]

$$\psi^{[A_2]}_{\zeta_1,\zeta_2}(\mathbf{k}, \mathbf{R}_1, \mathbf{R}_2) = \sin(k_x(X_1 - X_2)) \times$$
$$\times \frac{1}{\sqrt{2}}\left[u^*(\mathbf{k},\zeta_1)\, u^*(-\mathbf{k},\zeta_2)\, e^{ik_y(Y_1-Y_2)} - u^*(\mathbf{k},\zeta_2)\, u^*(-\mathbf{k},\zeta_1)\, e^{-ik_y(Y_1-Y_2)}\right] \chi \tag{3}$$

where the origin of the cell has been denoted with the capital $\mathbf{R} = (X, Y)$; χ_0 is the singlet spin function. We can verify by direct inspection that $\psi^{[A_2]}_{\zeta_1,\zeta_2}(\mathbf{k}, \mathbf{R}_1, \mathbf{R}_2)$ vanishes for $X_1 = X_2$, that is the two-body singlet wavefunction vanishes if the particles lie on the same annulus of the (N, N) tube. As a consequence $\psi^{[A_2]}_{\zeta_1,\zeta_2}(\mathbf{k}, \mathbf{R}_1, \mathbf{R}_2)$ is an eigenstate of the kinetic energy H_0 [with eigenvalue $2\varepsilon(\mathbf{k})$] and of the on-site Hubbard repulsion W with vanishing eigenvalue of the latter, that is $\psi^{[A_2]}_{\zeta_1,\zeta_2}(\mathbf{k}, \mathbf{R}_1, \mathbf{R}_2)$ is a $W = 0$ pair. The wave-function $\psi^{[A_2]}$ vanishes when the *transverse* component $k_y = 0$.

3 $W = 0$ Pairing: General Theory

In this Section we present our proposal for an electronic pairing mechanism. Looking at the experimental data we also argue that our description contains much of the essential physics. One of the distinct advantages of our approach over some competitors in the literature is that it lends itself to detailed comparison between the theory and the exact numerical data from clusters. The size of the clusters can be so small that we can comfortably afford exact calculations, but the choice of the geometry is important to ensure that the symmetry based theoretical formulation applies directly without misleading complications. Next, we show how this philosophy applies, using data from a tiny SWNT model and those from a square CuO_4 cluster in parallel. The latter geometry is relevant to Cu-O planes, and the way pairing is produced is similar. So we may claim that the $W = 0$ mechanism has enough universal character to be relevant in principle at least to other case besides SWNT.

We must define an appropriate effective interaction among the electrons of a $W = 0$ pair added to the n-body interacting ground state $|\Psi_0(n)\rangle$. Although the two extra particles cannot interact directly by definition of $W = 0$ pair, they can virtually exchange electron-hole excitations. To the extent that electron-hole excitations behave like Bose particles, they can act as the phonons do in the Frölich interaction. Incidentally, this is what all the electronic mechanisms of pairing on the market aim to do, but we place special emphasis on the role of symmetry, degeneracy and our special pairs.

Many configurations contribute to the interacting $(n+2)$-body ground state $|\Psi_0(n+2)\rangle$ and we need a complete set S to expand it exactly; as long as it is complete, however, we can design S as we please. We can take the non-interacting n-body Fermi *sphere* $|\Phi_0(n)\rangle$ as our vacuum and build the complete set in terms of excitations over it. In the subspace with vanishing spin z component, the simplest states that enter the configuration mixing are those obtained from $|\Phi_0(n)\rangle$ by creating two extra electrons over it; we denote with $|m\rangle$ these states. Similarly, along with the pair m states, we introduce the 4-body α states, obtained from $|\Phi_0(n)\rangle$ by creating 2 electrons and 1 electron-hole (e-h) pair. Then S includes the 6-body β states having 2 electrons and 2 e-h pairs, and so on. We are using Greek indices for the configurations containing the electron-hole pairs, which here are playing largely the same rôle as phonons in the Cooper theory. By means of the complet set S we now expand the interacting ground state

$$|\Psi_0(n+2)\rangle = \sum_m a_m|m\rangle + \sum_\alpha a_\alpha|\alpha\rangle + \sum_\beta a_\beta|\beta\rangle + ... \tag{4}$$

and set up the Schrödinger equation

$$H|\Psi_0(n+2)\rangle = E(n+2)|\Psi_0(n+2)\rangle. \tag{5}$$

We stress that Eq.(4) is configuration interaction, *not a perturbative expansion*. When the number n of electrons in the system is such that $|\Phi_0(n)\rangle$ is a single non-degenerate determinant (the Fermi surface is totally filled), we can easily and unambiguously define and calculate the effective interaction between the two extra electrons since the expansion in Eq.(4) for the interacting ground state is unique: this is done by a canonical transformation[14],[20],[21] from the many-body Hamiltonian of Eq.(1). We consider the effects of the operators H_0 and W on the terms of $|\Psi_0(n+2)\rangle$. Choosing the m, α, β, ... states to be eigenstates of the kinetic energy H_0 we have

$$H_0|m\rangle = E_m|m\rangle, \quad H_0|\alpha\rangle = E_\alpha|\alpha\rangle, \quad H_0|\beta\rangle = E_\beta|\beta\rangle, \quad \ldots. \tag{6}$$

Since W can create or destroy up to 2 e-h pairs, its action on an m state yields

$$W|m\rangle = \sum_{m'} W_{m',m}|m'\rangle + \sum_\alpha W_{\alpha,m}|\alpha\rangle + \sum_\beta W_{\beta,m}|\beta\rangle. \tag{7}$$

The action of W on the α states yields

$$W|\alpha\rangle = \sum_m W_{m,\alpha}|m\rangle + \sum_{\alpha'} W_{\alpha',\alpha}|\alpha'\rangle + \sum_\beta W_{\beta,\alpha}|\beta\rangle + \sum_\gamma W_{\gamma\alpha}|\gamma\rangle, \tag{8}$$

where scattering between 4-body states is allowed by the second term, and so on. In this way we obtain an algebraic system for the coefficients of the configuration interaction of Eq.(4). However to test the instability of the Fermi liquid towards pairing it is sufficient to study the amplitudes a_m of the m states. In the weak coupling limit this can be done by truncating the expansion in Eq.(4) to the α states because, as we have shown[20], the inclusion of the β, γ, ... states produces a E-dependent renormalization of the matrix elements of higher order in W, leaving the structure of the equations unaltered.

By taking a linear combination of the α states in such a way that

$$(H_0 + W)_{\alpha,\alpha'} = \delta_{\alpha\alpha'} E'_\alpha \tag{9}$$

the algebraic system reduces to

$$[E_m - E(n+2)]\, a_m + \sum_{m'} a_{m'} W_{m,m'} + \sum_\alpha a_\alpha W_{m,\alpha} = 0 \tag{10}$$

$$[E'_\alpha - E(n+2)]\, a_\alpha + \sum_{m'} a_{m'} W_{\alpha,m'} = 0. \tag{11}$$

Solving for a_α and substituting in Eq.(10) we exactly decouple the 4-body states as well, ending up with an equation for the dressed pair $|\psi\rangle = \sum_m a_m|m\rangle$. The effective Schrödinger equation for the pair reads

$$(H_0 + W + S[E])\,|\psi\rangle \equiv H_{\text{pair}}|\psi\rangle = E|\psi\rangle \tag{12}$$

where

$$(S[E])_{m,m'} = -\sum_{\alpha} \frac{W_{m,\alpha} W_{\alpha,m'}}{E'_{\alpha} - E}. \tag{13}$$

is the scattering operator. The matrix elements $W_{m,m'}$ in Eq.(12) may be written as the sum of two terms representing the direct interaction $W^{(d)}_{m,m'}$ among the particles forming the pair and the first-order self-energy W_m:

$$W_{m,m'} = W^{(d)}_{m,m'} + \delta_{m,m'} W_m. \tag{14}$$

Analogously in $S[E]$ we may recognize two different contributions; one is the true effective interaction W_{eff} between the electrons of the m states, while the other one is the forward scattering term F

$$S_{m,m'} = (W_{\text{eff}})_{m,m'} + F_m \delta_{m,m'} . \tag{15}$$

The first-order self-energy and the forward scattering term are diagonal in the indices m and m'. W_m and F_m renormalize the non-interacting energy E_m of the m states:

$$E_m \to E^{(R)}_m = E_m + W_m + F_m. \tag{16}$$

Eq.(12) is of the form of a Schrödinger equation with eigenvalue $E(n+2)$ for the added pair with the interaction $W^{(d)} + W_{\text{eff}}$. Here the $W = 0$ pairs are special because $W^{(d)}$ vanishes. We interpret a_m as the wave function of the dressed pair, which is acted upon by an effective Hamiltonian H_{pair}. This way of looking at Eq.(12) is perfectly consistent, despite the presence of the many-body eigenvalue $E(n+2)$. Indeed, if the interaction is attractive and produces bound states the spectrum of Eq.(12) contains discrete states below the threshold of the continuum (two-electron Fermi energy). This is a clear-cut criterion for pairing, which is exact in principle. The threshold is given by

$$E^{(R)}_F \equiv \min_{\{m\}}[E^{(R)}_m(E)], \tag{17}$$

which contains all the pairwise interactions except those between the particles in the pair; it must be determined once Eq.(12) has been solved (since F depends on the solution). The ground state energy E may be conveniently written as $E^{(R)}_F + \Delta$. $\Delta < 0$ indicates a Cooper-like instability of the normal Fermi liquid and its magnitude represents the binding energy of the pair.

We emphasize that in principle the canonical transformation is exact because in this way our framework does not require U/t to be small. The next problem is how to find a practical estimate of the renormalized Fermi energy. In the numerical calculations, some approximation is needed. In Sections 4 and 5, we shall compute the *bare* quantities; that is, we shall neglect the 6-body and higher excitations in the calculation of W_{eff} and F. This is a good approximation if the corrections to the Fermi liquid background are small and the exact numerical results on symmetric clusters suggest that this is the case, see next Section; the canonical transformation can be pushed to include more terms if necessary.

We want to stress that once the expansion of $|\Psi_0(n+2)\rangle$ in Eq.(4) is truncated as specified above we do not need to construct a good approximation of the interacting ground state wave function in order to get the a_m amplitudes at weak coupling; in this way we obtain information about pairing.

We conclude this Section by stressing that the Hubbard model is an instructive oversimplification of reality. Therefore it would be appropriate to include the effects of phonons within the theory described above. In this case, the $W = 0$ mechanism could still be very important as a way to get rid of the huge on-site repulsion; but then it could be unable to provide enough binding energy without *allies*. This scenario will be investigated in Section 4.1.

4 Results on Symmetric Clusters: Carbon Nanotubes vs Cuprates

The canonical transformation outlined above is completely general and can be readly applied to any system having enough symetry for $W = 0$ pairs to occur. In this Section we show that $W = 0$ pairing realizes both in hexagonal and square systems, suggesting that superconductivity in the wrapped graphite sheets of carbon nanotubes and in the Cu-O planes of cuprates can in principle have a common origin.

We shall consider two prototype clusters, namely the the $(1, 1)$ tube of length $L = 2$ (in units of the lattice spacing) and periodic boundary conditions and the CuO_4 plaquette, see Fig.(2). These systems are the smallest ones with the appropriate symmetry showing the superconducting $W = 0$ pairing. Therefore, they represent a very good probe to test the pairing mechanism shown in previous Section, since we can compare exact diagonalization results with the analytic approximations of the canonical transformation.

We define, following Refs.[22][23],

$$\tilde{\Delta}(\mathcal{N}+2) = E(\mathcal{N}+2) + E(\mathcal{N}) - 2E(\mathcal{N}+1), \tag{18}$$

where $E(\mathcal{N})$ is the ground state energy with $\mathcal{N}$ electrons (holes for Cuprates) referenced to the electron vacuum. $|\tilde{\Delta}(\mathcal{N}+2)|$ is one definition of the pairing energy. This definition is simple, but requires computing the eigenvalues with great accuracy, and has several drawbacks. It says nothing about the dynamics which leads to pairing. Moreover, generally a negative $\tilde{\Delta}$ does not unambiguously imply pairing, and further problems arise since the above definition depends on the comparison of systems with different $\mathcal{N}$. However, in several studies of $W = 0$ pairing in finite systems when it was possible to compute $\tilde{\Delta}$ by exact diagonalization we pointed out[13][15][21] that at least at weak coupling it agrees well with Δ as obtained by the canonical transformation. This supports the application of Eq.(18).

Below we compute the interacting ground state energy with 2, 3 and 4 particles by exact numerical diagonalization in order to get $\tilde{\Delta}(4)$ for the (1,1) nanotube and for the CuO_4 cluster. Finally the canonical transformation is applied to evaluate $\Delta(4)$ which will be compared with $\tilde{\Delta}(4)$.

For the clusters depicted in Fig.(2), we have determined the Optimal Group[10][24] and found that the low energy one-body spectrum consists of a non-degenerate foundamental level and a threefold degenerate first-excited one. By applying the Theorem in Section 2.1,

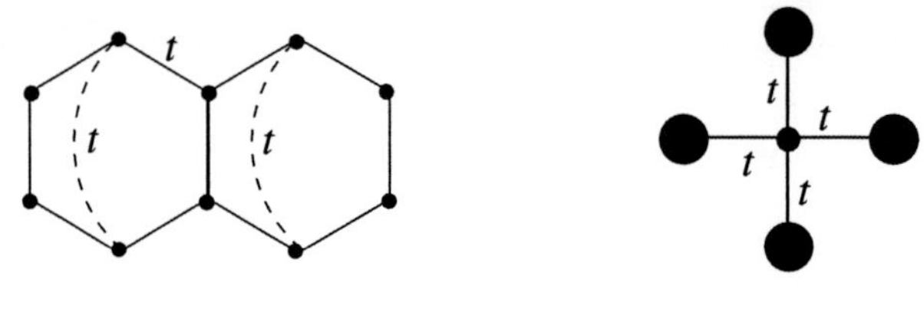

Figure 2: (a) The Hubbard Model of the (1,1) nanotube of length $L = 2$ (in units of the lattice spacing) and periodic boundary conditions; the dashed lines denote hopping interactions due to the wrapping. (b) CuO_4 cluster.

$W = 0$ pairs were found in both clusters and hence the possibility of negative Δ and $\tilde{\Delta}$ can be investigated.

In the non-interacting limit the ground state with 4 fermions is obtained by adding a $W = 0$ pair to the Fermi *sphere*, i.e. the non-interacting ground state with two fermions. In Fig.(3) $\tilde{\Delta}(4)$ is reported for the two cases and it is negative for a large range of the ratio U/t. In this range the effective interaction between the particles of the added $W = 0$ pair is *attractive*. The similarity of the curves in Fig.(3) suggests that the paring mechanism in the two prototype systems is actually the same. Hence, the $(1, 1)$ tube results to be a very good probe to test the canonical transformation for armchair carbon nanotubes as the CuO_4 is for Cuprates.

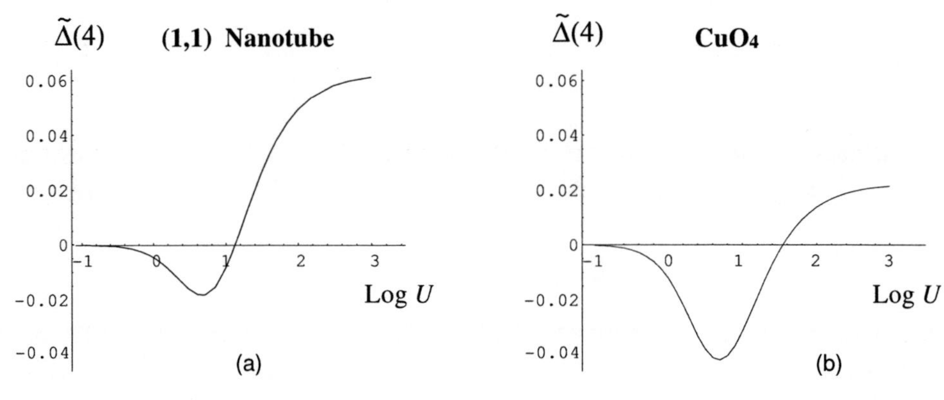

Figure 3: Trend of $\tilde{\Delta}(4)$ versus Log U in the range -1÷3 (a) for the (1,1) nanotube and (b) for the CuO_4. Energies are in unit of t.

The analytic canonical transformation can be readly done and the results are shown in Fig.(4). The agreement between $\tilde{\Delta}(4)$ and Δ is again very good. The ratio of the second derivatives with respect to U/t at $U = 0$ was estimated by using best fits and turned out to be 1.00003, while the first derivative vanishes. The binding energy for $U < t$ is in the meV range.

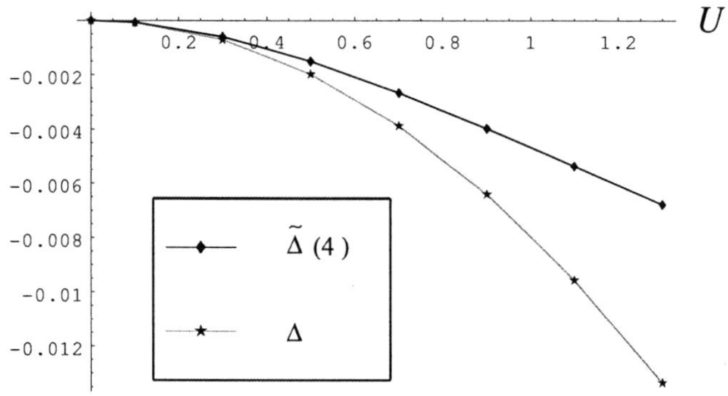

Figure 4: Comparison between $\tilde{\Delta}(4)$ and its analytic counterpart Δ versus U for the (1,1) nanotube of length L=2; energies are in units of t.

4.1 $W = 0$ Pairing and Electron-Phonon Interactions

Many high-T_C compounds exhibit a measurable isotope effect, which depends on doping. This experimental fact, the partly ionic nature of the oxides, and the existence of phonon frequencies in the same range as the pair binding energy suggests that Electron-Phonon (EP) interactions could be important and should be included in the theory. In particular there is experimental evidence that the half breathing Cu-O bond stretching mode at $k = (\pi, 0), (0, \pi)$ is significantly coupled with the doped holes in the superconducting regime and its contribution may be relevant for the $d_{x^2-y^2}$ pairing[25]. Even the Hubbard model for the SWNT can be questioned on similar grounds. In the conventional BCS mechanism, phonons overscreen the electron repulsion; what happens if electronic screening already leads to pairing? It is not obvious that the phonons will reinforce the attraction while preserving the symmetry. More generally, some vibrations could favour the pairing and others could be pair-breaking. This poses a special, conceptual difficulty for the present mechanism since any distortion of the lattice can be fatal to the $W = 0$ pairs.

Therefore, we address the general question if the $W = 0$ pairing available in the Hubbard model survives when the lattice degrees of freedom are taken into account. We use an extension of the Hubbard model in which bond stretchings dictate the couplings to the normal modes of the C_{4v}-symmetric configuration. In this way a long-range (Fröhlich) EP interaction arises.

Consider the Hubbard model appropriate for the Cu-O plane with on-site interaction U. We expand the hopping integrals $t(\mathbf{r}_i, \mathbf{r}_j)$ in powers of the displacements ρ_i around a C_{4v}-symmetric equilibrium configuration

$$t(\mathbf{r}_i, \mathbf{r}_j) \simeq t^0(\mathbf{r}_i, \mathbf{r}_j) + \sum_\alpha \left[\frac{\partial t(\mathbf{r}_i, \mathbf{r}_j)}{\partial r_i^\alpha}\right]_0 \rho_i^\alpha + \sum_\alpha \left[\frac{\partial t(\mathbf{r}_i, \mathbf{r}_j)}{\partial r_j^\alpha}\right]_0 \rho_j^\alpha,$$

where $\alpha = x, y$. Below, we write down the ρ_i^α in terms of the normal modes $q_{\eta\nu}$: $\rho_i^\alpha = \sum_{\eta\nu} S_{\eta\nu}^\alpha(i)\ q_{\eta\nu}$, where η is the label of an irreducible representation of the symmetry group of the undistorted system and ν is a phonon branch.

The Cu atoms is fixed, for the sake of simplicity; the electron-lattice Hamiltonian reads:

$$H_{el-latt} = H_0 + V_{\text{tot}} ,$$

where $H_0 = H_0^n + H_0^e$ accounts for the kinetic part of the electron-phonon system. We have

$$H_0 = \sum_{\eta} \hbar\omega_{\eta,\nu} b^\dagger_{\eta,\nu} b_{\eta,\nu} + \sum_{i,j\sigma} t^0(\mathbf{r}_i, \mathbf{r}_j)\, c^\dagger_{i\sigma} c_{j\sigma},$$

where the ω's are normal mode frequencies. The interacting part $V_{\text{tot}} = V + W$ contains the Hubbard repulsion W and the electron-phonon interaction V. The latter can be written in terms of the parameters $\xi_{\eta,\nu} = \lambda_{\eta\,\nu}\sqrt{\frac{\hbar}{2M\omega_{\eta,\nu}}}$, where the λ's are numbers of order unity that modulate the EP coupling strength, while M is the Oxygen mass. Then,

$$V = \sum_{\eta,\nu} \xi_{\eta,\nu}(b^\dagger_{\eta,\nu} + b_{\eta,\nu}) H_{\eta,\nu},$$

and the $H_{\eta,\nu}$ operators are given by

$$H_{\eta,\nu} = \sum_{i,j}\sum_{\alpha,\sigma} \left\{ S^{\alpha}_{\eta\,\nu}(i) \left[\frac{\partial t(\mathbf{r}_i,\mathbf{r}_j)}{\partial r_i^{\alpha}}\right]_0 + S^{\alpha}_{\eta\,\nu}(j) \left[\frac{\partial t(\mathbf{r}_i,\mathbf{r}_j)}{\partial r_j^{\alpha}}\right]_0 \right\} (c^\dagger_{i,\sigma} c_{j,\sigma} + h.c.)\,.$$

This is physically more detailed than the Hubbard-Holstein model, where electrons are coupled to a local Einstein phonon and the superconducting phase has been investigated in detail[26],[27]. Indeed, in the present context the Hubbard-Holstein model is not fully satisfactory because it restricts to on-site EP couplings, which would be impaired by a strong Hubbard repulsion. On the other hand the use of a Fröhlich-like phonons was suggested[28],[29],[30] for modeling the Cu-O planes in the strong EP coupling regime: a long-range EP coupling removes the problem of polaron self-trapping, otherwise present in the case of the Holstein interaction, where an unphysically large polaron (and bipolaron) mass occurs.

By generalizing to $H_{el-latt}$ the canonical transformation proposed for the electronic part $H = H_0^e + W$, one can derive[31],[32] an effective interaction between the particles in the pair. We obtained a new Cooper-like equation $H_{\text{pair}}|\varphi\rangle = E|\varphi\rangle$ with an effective two-body Hamiltonian, acting upon the dressed $W = 0$ pair $|\varphi\rangle$. As in Section 3 the pairing criterion involves the properly renormalized Fermi energy $E_F^{(R)}$: if the lowest energy eigenvalue E is such that $E = E_F^{(R)} + \Delta$ with negative Δ, the dressed $W = 0$ pair gets bound in the many-body interacting problem and the system undergoes a Cooper instability. We observed that this extended approach is very accurate at weak coupling and is qualitatively predictive also in the intermediate coupling regime.

The numerical calculations were done on CuO_4, which yields $\tilde{\Delta}(4) < 0$. The cluster CuO_4 is actually so small that it allows only the coupling to phonons at the centre or at the edge of the Brillouin Zone. Fortunately, however, phonons near the edge are precisely the most important ones for this system[25]. A standard Jahn-Teller calculation in which the degenerate ground state wave functions interact with the vibrations predicts distortions that already at moderate Electron-Phonon coupling destroy the pairing[32]. We also explored the scenario beyond this approximation. We have solved analytically the Cooper-like equation and found instead that the phonons help the electronic pairing to occur.

We resorted to exact numerical solutions in order to go beyond the weak-coupling regime, although we still used mild enough couplings to achieve the convergence with few excitations per mode. It turned out that the catastrophic distortions predicted by the Jahn-Teller approximation are an artifact, and that one needs a a broader spectrum of electronic states than is usually considered to get the correct answer. We could also explore the strong coupling regime by discarding the less influential modes and by including many excitations for the relevant modes. Interestingly, in the symmetry channels where $W = 0$ pairs occur, pairing may survive. We praise the fact that the numerics validate the results of the analytical canonical transformation. The analytical calculation also correctly perdicts which normal modes are pair-hailing and which are pair-breaking. In summary, our findings suggest a joint mechanism, with the Hubbard model that captures a crucial part of the physics.

5 Pairing in Doped Carbon Nanotubes

The results of the previous Section encourage us to proceed with larger and more physical systems than the $(1, 1)$ SWNT. In this Section we calculate the effective interaction [whose diagrammatic expansion is depicted in Fig.(5)] between the electrons of the $W = 0$ pair described by Eq.(3) in (N, N) nanotubes of length L with $4 < N < 16$ and $10 < L < 32$ (in units of the lattice spacing).

Following the general theory described in Section 3, we calculate the quantities entering the Cooper Equation (12). Letting $n_\nu^{(0)}(\mathbf{p})$ denote the non-interacting occupation number

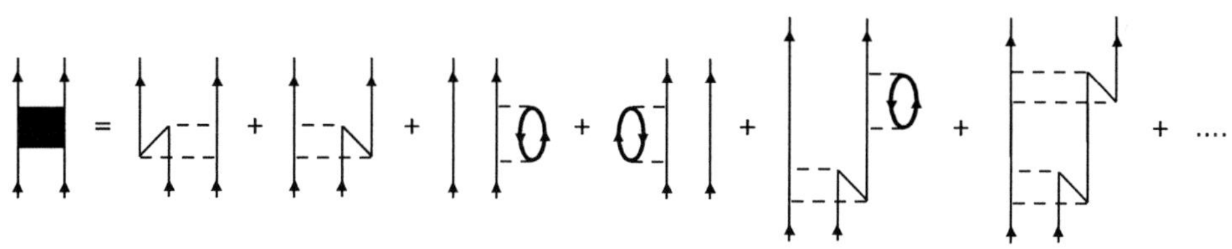

Figure 5: The effective vertex in H_{pair} (blob) expanded in terms of fermion lines (solid) and two-body interactions (dashed). Linked diagrams contribute to W_{eff}, unlinked ones to F.

in band ν with wavevector $\mathbf{p}$, we find

$$
\begin{aligned}
W_{\text{eff}}(\mathbf{k}, \mathbf{k}', E) &= 2 \sum_{\hat{O} \in C_{2v}} \chi^{(A_2)}(\hat{O}) \sum_{\mathbf{p}, \nu} [1 - n_+^{(0)}(\hat{O}\mathbf{k}' + \mathbf{k} + \mathbf{p})]\, n_\nu^{(0)}(\mathbf{p}) \times \\
&\times \frac{U_\nu(\hat{O}\mathbf{k}' + \mathbf{k} + \mathbf{p}, -\mathbf{k}, \hat{O}\mathbf{k}', \mathbf{p})\, U_\nu(\mathbf{k}, \mathbf{p}, \hat{O}\mathbf{k}' + \mathbf{k} + \mathbf{p}, -\hat{O}\mathbf{k}')}{\varepsilon^+(\hat{O}\mathbf{k}' + \mathbf{k} + \mathbf{p}) - \varepsilon^\nu(\mathbf{p}) + \varepsilon^+(\mathbf{k}') + \varepsilon^+(\mathbf{k}) - E} \qquad (19)
\end{aligned}
$$

where $\chi^{(\eta)}(\hat{O})$ is the character in η of the operation $\hat{O}$ of C_{2v}, E is the interacting pair energy and $U_\nu(\mathbf{k}_1, \mathbf{k}_2, \mathbf{k}_3, \mathbf{k}_4)$ is the interaction vertex, with incoming legs $\mathbf{k}_3$ and $\mathbf{k}_4$ in band $+$ and outgoing $\mathbf{k}_1$ in band $+$ and $\mathbf{k}_2$ legs in band ν. The effective Schrödinger equation

Table 1: Pair binding energy $-\Delta$ (meV) and Average Effective Interaction V (eV) for (N, N) nanotubes of various lengths L, as a function of the Fermi energy ε_F. Numerical values were computed with $t = 1$ eV and $U = 1.7$ eV for illustration.

L	$N=2$					$N=4$					$N=6$				
10	ε_F	0.5	0.8	1.0	1.2	ε_F	0.5	0.8	1.0	1.2	ε_F	0.5	0.8	1.0	1.2
	$-\Delta$	12.0	11.6	11.1	10.5	$-\Delta$	6.3	6.0	6.5	5.5	$-\Delta$	4.0	4.2	4.4	4.4
	$-V$	0.43	0.42	0.40	0.40	$-V$	0.37	0.40	0.24	0.40	$-V$	0.39	0.38	0.17	0.25
20	ε_F	0.5	0.8	1.0	1.2	ε_F	0.5	0.8	1.0	1.2	ε_F	0.5	0.8	1.0	1.2
	$-\Delta$	5.7	6.0	6.0	6.0	$-\Delta$	3.1	3.0	2.9	3.2	$-\Delta$	2.4	2.2	3.5	1.6
	$-V$	0.38	0.40	0.40	0.44	$-V$	0.38	0.40	0.15	0.43	$-V$	0.47	0.36	0.25	0.34
32	ε_F	0.5	0.8	1.0	1.2	ε_F	0.5	0.8	1.0	1.2	ε_F	0.5	0.8	1.0	1.2
	$-\Delta$	3.6	3.7	3.9	3.9	$-\Delta$	2.0	2.0	1.8	1.9	$-\Delta$	1.5	1.1	1.6	1.0
	$-V$	0.35	0.36	0.38	0.45	$-V$	0.35	0.43	0.14	0.37	$-V$	0.38	0.30	0.18	0.33

for the pair reads

$$[2\varepsilon(\mathbf{k}) + W_F + F(\mathbf{k}, E)]\, a_{\mathbf{k}} + \sum_{\mathbf{k}' \in \mathcal{D}/4} W_{\text{eff}}\,(\mathbf{k}, \mathbf{k}', E)\, a_{\mathbf{k}'} = E a_{\mathbf{k}}\ , \tag{20}$$

where W_F is the $\mathbf{k}$-independent first-order self-energy shift [see Eq.(14)] given by

$$W_F = U\left[2\left(\frac{n/2}{4NL}\right)^2 + \frac{n/2}{4N^2L^2}\right] \tag{21}$$

and

$$F(\mathbf{k}, \mathbf{k}', E) = -2\delta(\mathbf{k}-\mathbf{k}') \times$$
$$\sum_{\mathbf{p},\nu}\sum_{\mathbf{q}} \frac{[1 - n_+^{(0)}(\mathbf{k}+\mathbf{p}-\mathbf{q})]\,[1 - n_+^{(0)}(\mathbf{q})]\, n_\nu^{(0)}(\mathbf{p})}{\varepsilon^+(\mathbf{k}+\mathbf{p}+\mathbf{q}) - \varepsilon^\nu(\mathbf{p}) + \varepsilon^+(\mathbf{q}) + \varepsilon^+(\mathbf{k}) - E}\, |U_\nu(\mathbf{k}, \mathbf{p}, \mathbf{k}+\mathbf{p}-\mathbf{q}, \mathbf{q})|^2 \tag{22}$$

is the *forward scattering* term which does not contains any direct interaction between the particles of the pair. Eq.(20) requires a self-consistent calculation of E (since W_{eff} and F are E-dependent). The indices $\mathbf{k}$ and $\mathbf{k}'$ run over $1/4$ of the empty part of the First Brillouin Zone and we denoted such a set of wavevectors as $\mathcal{D}/4$. We show below that $E = 2\varepsilon_F + W_F + F_{\min} + \Delta$, with $\Delta < 0$ which implies a positive binding energy $-\Delta$ of the $W = 0$ pair; here $F_{\min}$ is the minimum value of $F(\mathbf{k}, E)$ among the wavevectors $\mathbf{k}$ on the Fermi line.

We solved the Cooper-like equation (20) in a virtually exact way using the hopping parameter $t = 1$ eV and $U/t = 1.7$ (which is of the correct order of magnitude for graphite [33][34]). The calculations are performed with the Fermi energy ε_F varying between 0.5 eV and 1.2 eV (half filling corresponds to $\varepsilon_F = 0$). As in the $(1, 1)$ cluster, the $W = 0$ singlets show pairing, albeit in general in ${}^1A_2$, see Table I. The binding energy $-\Delta$ of the pairs decreases monotonically both with the radius and the length of the tube.

With supercell sizes $N_C = 2N \times L > 400$ numerical calculations become hard. Since we are concerned with the asymptotic behaviour for fixed N and $L \to \infty$ and $\Delta(N, L)$ depends on N and L in a complicated way, we need a method to make reliable extrapolations of the numerical results. To this end, like in previous work[14],[35] we define the Average

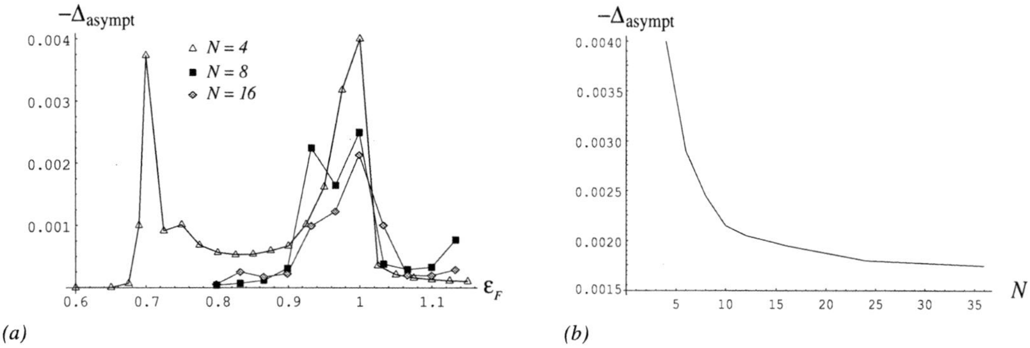

Figure 6: Results of the canonical transformation approach with $t = 1$ eV and $U = 1.7$ eV. (a) $-\Delta_{\text{asympt}}$ (eV) as a function of the Fermi energy ε_F for $N = 4$ (empty triangles), $N = 8$ (black boxes) and $N = 16$ (grey diamonds). The Fermi energy varies in the range $0.6 \div 1.2$ eV. (b) $-\Delta_{\text{asympt}}$ (eV) as a function of N for N in the range $4 \div 36$ and $\varepsilon_F = 1$ eV.

Effective Interaction V. This is such that setting in Eq.(20) $W_{\text{eff}} = -\frac{V}{N_C}$, with a constant $V > 0$ for all $\mathbf{k}$ and $\mathbf{k}'$ in $\mathcal{D}/4$, one obtains the correct value of Δ, i.e., the same as obtained by solving Eq. (20) with $W_{\text{eff}}(\mathbf{k}, \mathbf{k}')$. In other terms, once the binding energy $-\Delta(N, L)$ is known, the constant V must be chosen in such a way that

$$\frac{1}{V} = \frac{1}{N_C} \sum_{\mathbf{k} \in \mathcal{D}/4} \frac{1}{[2\varepsilon(\mathbf{k}) + F(\mathbf{k})] - [2\varepsilon_F + F_{\min}(\mathbf{k}_F)] - \Delta(N, L)}. \tag{23}$$

In Table I we report the value of V which remains fairly stable around ≈ 0.4 eV with increasing L. Therefore V must be interpreted as a characteristic energy scale of the system which is largely independent of the Fermi energy and of the radius. This allows us to extrapolate to $\Delta_{\text{asympt}}(N) = \lim_{L \to \infty} \Delta(N, L)$. Solving Eq.(23) with $V \approx 0.4$ eV for $L \to \infty$ the new unknown $\Delta_{\text{asympt}}(N)$ can be obtained numerically. We did so for several values of N and ε_F and found that Δ_{asympt} is strongly dependent on the filling at fixed N, as illustrated in Fig.(6.a). The sharp maximum at the *optimal doping* $\varepsilon_F \approx 1$ can be understood in terms of a corresponding peak in the density of states; we observe that $-\Delta_{\text{asympt}}(N)$ decreases monotonically as the radius of the tube increases [see Fig.(6.b)] *i.e.* a higher critical Temperature is predicted for smaller nanotube radii. This is also supported by the measurements of a $T_c \approx 15$ K in the 4 Ångstrom SWNT by Tang *et al.*[5]. However, in the limit of large N, $\Delta_{\text{asympt}}(N)$ remains stable around 1.7 meV and may be interpreted as the binding energy of the $W = 0$ pair in an *optimally doped* graphite sheet. By a rough order of magnitude estimate, we may say that the superconducting critical temperature predicted by our approach at $\varepsilon_F = 1$ eV (which corresponds to a number of electrons per graphite atom of 1.25) is $T_c \approx \Delta_{\text{asympt}}(\infty) \approx 10 \div 20$ K. The order of magnitude of the computed pair binding energy agrees with experiment[36].

6 Conclusions

For (N, N) armchair nanotubes, the effective interaction W_{eff} of Eq.(19) arises from the on-site electronic repulsion. Inserting W_{eff} in Eq.(20) leads to pairing and the binding energy is in the meV range. Despite the heuristic nature of the Hubbard Model, it is gratifying that the order of magnitude of the pair binding energy agrees well with experiment[5]. Furthermore, the decrease of the binding energy with N is suggested by recent measurements on nanotubes with diameter of few $\mathring{A}$ngstrom[5]. The analytical approximation has been validated with high accuracy by comparison with exact-diagonalization data in the case $N = 1$.

The paired state we have obtained here is essentially two-dimensional, that is the transverse direction is crucial to have a non-Abelian symmetry group and hence $W = 0$ pairs away from half filling. It has been argued that weakly doped nanotubes are Luttinger liquids down to milli-Kelvin temperatures[3][37]. This opens up an interesting possibility. In nanotubes, two distinct superconducting order parameters could appear in the phase diagram: in doped nanotubes, pairing occurs by the $W = 0$ scenario, while close to half-filling there could be another one due to a breakdown of the Luttinger liquid[38].

Carbon nanotubes and high-T_c Cuprates are widely different for composition and physical properties. However, symmetry arguments based on the above $W = 0$ theorem suggest that, despite the obvious differences, part of the story must be the same.

The present treatment based on a Hubbard model ignores vibrations from the outset and one should seek some way to include them in the model with the resulting possibility that static or dynamic Jahn-Teller distortions modify the problem. Here we reported some preliminary results obtained on a simple square lattice and compared to exact diagonalization data on a tiny (5-site) system. The results indicate that the situation is complex, as can be expected in an enlarged parameter space. However, exact solutions confirm that the $W = 0$ pairing survives and also gains stability, as opposed to those findings where simpler approximations would predict distortion and pair breaking. This is because the neglect of the excited states underestimates the flexibility of the pair ad its ability to *ride* the vibrations.

References

[1] S. Iijima, *Nature* (London) **354**, 56 (1991).

[2] M. Bockrath, D. H. Cobden, J. Lu, A. G. Rinzler, R. E. Smalley, L. Balents and P. L. McEuen, *Nature* (London) **397**, 598 (1999); Z. Yao, H. W. C. Postma, L. Balents and C. Dekker, *Nature* (London) **402**, 271 (1999); H. W. C. Postma, M. de Jonge, Z. Yao and C. Dekker, *Phys. Rev.* **B 62**, R10653 (2000).

[3] Jenö Sólyom, in *Proceedigs of the School and Workshop on Nanotubes and Nanostructures 2000*, Santa Margherita di Pula (Ca), Italy.

[4] A. F. Morpurgo, J. Kong, C. M. Marcus and H. Dai *Science* **286**, 263 (1999).

[5] Z. K. Tang, Lingyum Zhang, N. Wang, X. X. Zhang, G. H. Wen, G. D. Lee, J. N. Wang, C. T. Chan and Ping Sheng, *Science* **292**, 2462 (2001).

[6] M. Kociak, A. Yu. Kasumov, S. Gue'ron, B. Reulet, I. I. Khodos, Yu. B. Gorbatov, V. T. Volkov, L. Vaccarini and H. Bouchiat, *Phys. Rev. Lett.* **86**, 2416 (2001).

[7] A. Y. Kasumov, R. Deblock, M. Kociak, B. Reulet, H. Boukiat, I. I. Khodos, Yu. B. Gorbatov, V. T. Volkov, C. Journet and M. Burghard, *Science* **284**, 1508 (1999); see also A. Yu. Kasumov, in *Proceedigs of the School and Workshop on Nanotubes and Nanostructures 2000*, Santa Margherita di Pula (Ca), Italy.

[8] B. Gao, A. Kleinhammes, X. P. Tang, C. Bower, L. Fleming, Y. Wu and O. Zhou, *Chem. Phys. Lett.* **307**, 153 (1999).

[9] J. Zhao, A. Buldum, J. Han and J. Ping Lu, *Phys. Rev. Lett.* **85**, 1706 (2000).

[10] Enrico Perfetto, Gianluca Stefanucci and Michele Cini, *Phys. Rev.* **B 66**, 165434 (2002).

[11] Enrico Perfetto, Gianluca Stefanucci and Michele Cini, *Eur. Phys. J.* **B 30**, 139 (2002).

[12] S. Bellucci P. Onorato and J. Gonzalez, *Phys. Rev.* **B 69**, 085404 (2004).

[13] Michele Cini and Adalberto Balzarotti, *Phys. Rev.* **B 56**, 14711 (1997).

[14] Michele Cini, Gianluca Stefanucci and Adalberto Balzarotti, *Eur. Phys. J.* **B 10**, 293 (1999).

[15] Michele Cini, Enrico Perfetto and Gianluca Stefanucci, *Eur. Phys. J.* **B 20**, 91 (2001).

[16] Michele Cini and Gianluca Stefanucci, *Solid State Commun.* **117**, 451 (2001).

[17] Gianluca Stefanucci and Michele Cini, *J. Phys.: Condens. Matter* **14**, 2653 (2002).

[18] Gianluca Stefanucci and Michele Cini, *Phys. Rev.* **B 66**, 115108 (2002).

[19] Michele Cini, Adalberto Balzarotti, Raffaella Brunetti, Maria Gimelli and Gianluca Stefanucci, *Int. J. Mod. Phys.* **B 14**, 2994 (2000).

[20] Michele Cini, Gianluca Stefanucci and A. Balzarotti, *Solid State Commun.* **109**, 229 (1999).

[21] Michele Cini, Adalberto Balzarotti and Gianluca Stefanucci, *Eur. Phys. J.* **B14**, 269 (2000).

[22] J. E. Hirsch, S. Tang, E. Loh and D. J. Scalapino, *Phys. Rev. Lett.* **60**, 1668 (1988).

[23] C. A. Balseiro, A. G. Rojo, E. R. Gagliano and A. Alascio, *Phys. Rev.* **B 38**, 9315 (1988).

[24] Adalberto Balzarotti, Michele Cini, Enrico Perfetto, and Gianluca Stefanucci, accepted for publication in *J. Phys.: Condens. Matter* (2004).

[25] Z. X. Shen, A. Lanzara, S. Ishihara, and N. Nagaosa, *Phil. Magazine* **B 82**, 1349-1368 (2002).

[26] C. H. Pao and H. B. Schuttler, *Phys. Rev.* **B 57**, 5051 (1998); *Phys. Rev.* **B 60**, 1283 (1999).

[27] A. S. Alexandrov, *Phys. Rev.* **B 46**, 2838 (1992); G. Wellein, H. Roder and H. Fehske, *Phys. Rev.* **B 53**, 9666 (1996); J. Bonca, T. Katrasnik and S.A. Trugman, *Phys. Rev. Lett.* **84**, 3153 (2000).

[28] A. S. Alexandrov and P. E. Kornilovitch, *J. Phys.* **C 14**,5337 (2002).

[29] A. S. Alexandrov, *Int. J. Mod. Phys.* **B 17**, 3315 (2003).

[30] J. Bonca and S. A. Trugman, *Phys. Rev.* **B 64**, 094507 (2001).

[31] E. Perfetto and M. Cini, *Phys. Rev.* **B 69**, 92508 (2004).

[32] E. Perfetto and M. Cini, *J. Phys.* **C 16** 4845 (2004).

[33] M. P. Lopez Sancho, M. C. Munoz and L. Chico, *Phys. Rev.* **B 63**, 165419 (2001).

[34] A. L. Tchougreef and R. Hoffman, *J. Phys. Chem.* **96**, 8993 (1992).

[35] W. Fettes and I. Morgenstern, *Eur. Phys. J.* **B 9**, 9509 (1996).

[36] J. T. Belash, O. V. Zharikov and A. V. Pal'nichenko, *Solid State Comm.* **63**, 153 (1987); J. T. Belash, A. D. Bronnikov, O. V. Zharikov and A. V. Pal'nichenko, *Solid State Comm.* **64**, 1445 (1987); J. T. Belash, A. D. Bronnikov, O. V. Zharikov and A. V. Pal'nichenko, *Solid State Comm.* **69**, 921 (1989).

[37] L. Balents and M. P. A. Fisher, *Phys. Rev.* **B 55**, R 11973 (1997).

[38] Y. A. Krotov, D. H. Lee and S. G. Louie, *Phys. Rev. Lett.* **78**, 4245 (1997).

In: Focus on Nanotube Research
Editor: Delores A. Martin, pp. 37-65

ISBN: 1-59454-851-X

Chapter 3

CLUSTER NATURE OF THE SOLVATION FEATURES OF SINGLE-WALL CARBON NANOTUBES

Francisco Torrens and Gloria Castellano
Institut Universitari de Ciència Molecular, Universitat de València,
Dr. Moliner-50, E-46100 Burjassot (València),
and Facultad de Ciencias Experimentales,
Universidad Católica de Valencia *San Vicente Mártir*,
Guillem de Castro-106, E-46003 València, Spain

Abstract

It is discussed the possibility of the existence of single-wall carbon nanotubes (SWNTs) in organic solvents in the form of clusters, containing a number of SWNTs. A theory is developed based on a bundlet model for clusters, which enables describing the distribution function of clusters by size. Comparison of the calculated values of solubility with experimental data would permit obtaining energetic parameters characterizing the interaction of an SWNT with its surrounding, in a solid phase or solution. Fullerenes and SWNTs are unique objects, whose behaviour in many physical situations is characterized by remarkable peculiarities. Peculiarities in solutions show up first in that fullerenes and SWNTs represent the only soluble forms of carbon. This is primary related to the originality in the molecular structure of fullerenes and SWNTs. The fullerene molecule is a virtually uniform closed spherical or spheroidal surface, having no sharp ridges or dents. Similarly, an SWNT is a smooth cylindrical unit. Both structures give rise to the relatively weak interaction between the neighbouring molecules in a crystal, and promote effective interaction of the molecules with those of a solvent. Another peculiarity in solutions is related to their tendency to form clusters, consisting of a number of fullerene molecules or SWNTs. The energy of interaction of a fullerene molecule or SWNT with solvent molecules is proportional to the surface of the former molecule, and roughly independent of the relative orientation of solvent molecules. All these phenomena have a unified explanation in the framework of the bundlet model of a cluster, in accordance with which the free energy of an SWNT involved in a cluster is combined from two components, *viz.* a volume one, which is proportional to the number of molecules n in a cluster, and a surface one, which is proportional to $n^{1/2}$. Both single- and complete-linkage cluster analyses (CAs) of the packing properties of equal objects allow classifying these properties and agree. Principal components analysis of the packing properties and CA of the packing objects permit classifying them and agree. The packing objects, clusters and other systems of atoms with short-range interaction, are grouped into three classes. The bundlet model for clusters enables describing the distribution function of SWNT clusters by size. A droplet model for clusters of conical single-wall carbon nanohorns (SWNHs) is also proposed. The droplet model is formally analogous to that for clusters of fullerenes. From purely geometrical differences the models predict different behaviours.

1 Introduction

Clusters (conglomerates of a number of atoms ranging from a few to several thousands) constitute a bridge between molecular systems and solids, and exhibit challenging features [1]. Frequently, an astonishingly large number of different configurations have similar energies, making it difficult to determine stable structures. Their melting properties can also be significantly different from those of the solid, due to the finite size, the presence of surfaces and the anisotropy. Metal clusters are extremely important from the technological point of view, due to their role as catalysts in important chemical reactions (*e.g.*, in catalytic exhaust pipes of cars). Clusters are important systems and simulation of clusters has been performed since the early days of Monte Carlo and molecular dynamics.

The study of atomic and molecular clusters is an active area of research in the field of nanosciences, both in the theoretical and experimental contexts. Nanoclusters essentially form a bridge between bulk materials and small molecules. A cluster can be considered as the bridge between molecules and solids. Clusters consist of a number of atoms ranging from a few to several thousands, encompassing interesting features. For example, a large number of different cluster configurations may have similar energies, thus making it difficult to determine which structure may be more stable. Clusters can also form metastable structures, which can transform into other stable configurations when energy is supplied. Melting temperatures also vary significantly from the macroscopic value or may not be determinable at all.

Due to the relative large number of surface atoms, a cluster exhibits different mechanical and chemical behaviours compared to a bulk material. Nanoclusters may offer a broad range of interesting possibilities not only in relation to material sciences, where they can be used to construct coating surfaces or new materials, but also in chemistry because of their high reactivity. Atom–cluster collision processes are also crucial in environmental and climate science problems. An example is homogeneous gas–liquid nucleation. In this case, gas–liquid nucleation results in the appearance of molecular clusters of supersaturated vapour. This essentially consists of a series of condensation and evaporation events where the clusters gain or lose molecules. In future nanoengineering applications, nanoclusters may serve as components of small devices. Since nanoscale phenomena deviate from the established laws and properties of the macroscopic world, one is well motivated to study the dynamic processes occurring on nanoscales, such as the impact of cluster onto surfaces and cluster–cluster collision.

The C atoms in fullerene molecules are located at the vertices of regular hexagons and pentagons, which cover in a regular manner the surface of a sphere or spheroid. The most spread and elaborated of the molecules belonging to the fullerene family is C_{60}, whose structure is a regular truncated icosahedron. The surface of this molecule is constructed from twenty regular hexagons and twelve regular pentagons so that each pentagon is adjacent only to hexagons, whereas each hexagon is adjacent to three pentagons and three hexagons alternately. The fullerene family also includes along with C_{60} the molecules C_{70}, C_{76}, C_{78}, C_{84}, etc., distinguished by a lower symmetry and larger number of hexagons on the surface. Thus, the fullerenes form a unique class of molecules having a closed two-dimensional structure.

The exotic structure of fullerenes also determines their extraordinary behaviour in solutions. Since the specific surface energies of interaction of fullerene molecules with one another and with solvent molecules are close in magnitude, fullerenes in solutions display a trend to the formation of aggregates or clusters consisting of a number of molecules. In a quite highly concentrated solution in the state of thermodynamic equilibrium, the overwhelming majority of dissolved substance is in the cluster form. This is the only example of a situation, when essentially the whole of substance is in cluster state. The situation, when the portion of substance involved in clusters is relatively low, seems a more typical. Thus, the study of the behaviour of fullerenes in solutions, not only provides new information on fullerene properties, but also extends our notions about macroscopic characteristics of a substance consisting mainly of clusters.

Extraordinary physical and chemical peculiarities in the behaviour of fullerenes in solutions are related, on the one hand, to their exotic structure and, on the other, to the possibility of cluster formation. This makes fullerene solutions an interesting object of chemical physics, possessing unusual thermodynamic, kinetic, optical and other properties. Among these unusual properties should be mentioned the nonmonotonic temperature dependence of solubility of fullerenes in some solvents [2] as well as the nonlinear concentration dependence of the third-order nonlinear optical susceptibility [3]. Of considerable scientific interest is the solvatochromic effect [4,5], which is exhibited in a sharp alteration in the spectrum of the optical absorption of C_{70}, dissolved in a mixture of organic solvents, of a result of a slight change in the solvent content. These and some other peculiarities in the behaviour of fullerenes in solutions are attributable to the phenomenon predicted theoretically and revealed in experiments [6–8] of the formation of clusters (consisting of a number of fullerene molecules) in a solution. A thermodynamic approach to the description of this phenomenon based on a droplet model of a cluster enables one uniquely to describe numerous peculiarities in the behaviour of fullerenes in solutions and, on this basis, to predict new effects such as the concentration and temperature dependences of the diffusion and thermal diffusion coefficients of fullerenes in solutions as well as the concentration dependence of the heat of solution of fullerenes.

It was examined the decrease in pyridine-soluble material observed after soaking coals in solvents, what is due to an increase in cross-linking associated with the formation of ionic domains or clusters, similar to those observed in a class of synthetic polymers known as ionomers [9].

Stable aqueous dispersions of fullerenes, C_{60-70}, were prepared by simply injecting into water a saturated solution of C_{60-70} in tetrahydrofuran (THF), followed by THF removal by purging gaseous nitrogen $N_{2(g)}$. It is not possible to extract C_{60-70} from a solution in toluene to water. However, it is also not possible to extract C_{60-70} from a dispersion in water to toluene. Similar results were reported for other C_{60-70} aqueous dispersion systems [10]. Upon contact with water, under a variety of conditions, C_{60} spontaneously forms a stable aggregate $(C_{60})_n$ with nanoscale dimensions (d = 25–500nm) [11]. The colour, hydrophobicity and reactivity of individual C_{60} are substantially altered in $(C_{60})_n$. It was shown that in solution $(C_{60})_n$ is crystalline in order, and remains as underivatized C_{60} throughout the formation/stabilization process that can later be chemically reversed. Particle size can be affected by formation parameters, *e.g.*, rates and the pH of the water addition. Once formed, $(C_{60})_n$ remains stable in solution at or below ionic strengths of 0.05mol·L^{-1} for months. Andrievsky *et al.* [12] and Deguchi *et al.* [13] suggested that water itself might form a donor–acceptor complex with C_{60} leading to a weakly charged colloid [14]. By contrast, C_{60}, dissolved in water *via* complexation with cyclodextrin$_8$ (γ-CD), was extracted to toluene [15,16]. In the case of C_{60} incorporated into artificial lipid membranes C_{60} was not extracted to toluene, but the extraction became possible once the vesicle was destructed by adding KCl [17]. Addition of KCl was also required to extract poly(vinylpyrrolidone)-solubilized $C_{60–70}$ to toluene [18]. When NaCl was added to a $C_{60–70}$ aqueous dispersion, $C_{60–70}$ were extracted into toluene and the toluene phase exhibited faint magenta or orange, characteristic colours for a solution of C_{60} or C_{70} in toluene, respectively. $C_{60–70}$ are dispersed as monodisperse clusters $(C_{60–70})_n$ in water, 60nm in diameter. High-resolution transmission electron microscopy revealed the polycrystalline nature of $(C_{60–70})_n$. The preparation of the dispersion is easy to perform, and the dispersions thus obtained are of excellent colloidal stability, even though no stabilizing agent is used. They found that the surface of $(C_{60–70})_n$ is negatively charged, and the electrostatic repulsion between the negatively charged $(C_{60–70})_n$ surfaces is important for the stability of the dispersions.

A *closest packing* is just another way to say *the densest packing* of objects in a room (*e.g.* in Tetris), useful when figuring out the best way to load the boot of a car. The closest packings of rigid spheres, like bowling balls, have gotten some special attention from scientists and mathematicians in the last few centuries. This is not because they were trying to figure out the most stable way to load their car's boots with bowling balls, but because rigid spheres are quite good approximations for little particles from dust grains to atoms. A lot of information, *e.g.* the rigidity of a metal alloy, can be predicted by treating these materials as a pile of differently sized bowling balls en miniature. One would not have to synthesize the alloy to see what happens. Notice that the *densest* packing is not necessarily the most *effective*, especially in quotidian life; *e.g.*, when loading a cubic box with spheres, an ordinary stack may allow putting more spheres into the box than the densest packing would. If there is not such a box, the densest packing is also the most effective and stable, *e.g.*, the well-known pile of cannonballs.

An assembly of randomly packed spheres can represent certain features of the geometry of simple liquids [19]. The density of random close packing, 0.637±0.005, corresponded closely to the densities of the rare-gas liquids at their triple points. The radial distribution of random close (RC) packing was found to agree well with the radial distributions of the rare-gas liguids determined by X-ray and neutron diffraction. The lower limit of the packing density measured was 0.601±0.005. Though random *loose* (RL) packing may not be as basic a geometrical arrangement as RC packing, it might correspond, *e.g.*, to a simple liquid at a temperature somewhat above its triple point. Models of randomly packed hard spheres exhibited some features of the properties of simple liquids, *e.g.* the packing density and the radial distribution [20]. The value of the maximum packing density of spheres can be determined from models, if care is taken to ensure random packing at the boundary surfaces, and if correction is made for volume errors at the boundaries. Experiments for both the random *loose* and the random close-packed densities were reported. A computer analysis of the results permitted a one-step, two-parameter extrapolation to infinite volume. The figure so obtained for the random close-packed density is 0.6366±0.0005, which represented an improvement in precision over previous results by an order of magnitude. Using a new series acceleration method, the virial expansion for the pressure of hard discs and hard spheres was found to be a monotonically increasing function of the number density ρ and diverged at the density of closest packing with the critical exponent $\gamma = 1$ [21]. The virial series described a metastable fluid phase for $\rho_0 < \rho < \rho_c$, where ρ_0 is the density of the fluid–solid transition. In the case of hard discs, the known values of the virial coefficients B_n were well represented by $B_n = (C - Dn^{-1})\rho_c^{-(n-1)}$ for $n \geq 4$, with C and D being positive constants. Adopting this approximation for $n \geq 7$ led to a simple, closed-form equation of state for the fluid phase, which yielded values of the pressure throughout the regime $0 < \rho < \rho_0$ that were only slightly larger than the series acceleration estimates.

The algorithm of packing non-congruent spheres was generalized for the case of packing arbitrary convex three-dimensional (3D) objects, suggesting an algorithm building close packing of arbitrary set of 3D special figures (*sphere-polyhedrons*) [22]. The model can be applied: (1) for investigation of properties of compact packings of complex form convex objects, *e.g.*, for studying influence of the form of objects onto the density and average of neighbours; (2) for creating starting configurations while structure–imitation modelling agglomeration and forming powders based on the particles method.

They were reviewed the calculations on solvents and co-solvents of single-wall carbon nanotubes (SWNTs) with cyclopyranoses [23], and partition of solvents and co-solvents of SWNTs with proteins and cyclopyranoses [24]. In earlier publications, periodic tables of SWNTs were discussed [25,26]. A program based on the AQUAFAC model was applied to calculate the aqueous coefficients of SWNTs [27]. A molecular modelling comparative study of SWNT solvents and co-solvents provided a classification in best, good and bad solvents [28–32]. The bundlet model for clusters of SWNTs was presented elsewhere [33–35]. The aim of the present report is to perform a comparative study of the properties of fullerenes (droplet model), and SWNTs (bundlet model).

In this article, a wide class of phenomena accompanying the behaviour of SWNTs in solutions is analyzed from a unique point of view, taking into account the tendency of SWNTs to cluster formation in solutions. The available experimental data are given concerning the solubility of fullerenes C_{60} and C_{70} in a variety of solvents at various temperatures, as well as the diffusion coefficient of fullerenes in solutions. Experiments revealing extraordinary peculiarities in the behaviour of fullerenes in solutions are analyzed. Those peculiarities are considered based on a droplet model of cluster, in which frames it is performed a consistent thermodynamic description of the behaviour of fullerenes in solutions. This provides a quantitative examination for observed unusual thermodynamic and optical characteristics of fullerene solutions, and the prediction of new effects in the solutions of fullerenes in organic solvents such as concentration and temperature dependences for diffusion and thermal diffusion coefficients, as well as concentration dependence for the heat of solution. Based on modern approaches to the description of mechanisms of fractal cluster growth, they are considered the experiments concerning the kinetics of fractal cluster formation in fullerene solutions. They are analyzed similarity laws determining the thermodynamic properties of fullerite crystals, which are characterized by the close-packed, face-centred cubic structure and short-range intermolecular interaction. Based on the droplet model for clusters of fullerenes, they are proposed the bundlet model for clusters of SWNTs, and the droplet model for clusters of single-wall carbon nanohorns (SWNHs). It is carried out a comparative study of the solubilities of fullerenes (droplet model) and SWNTs (bundlet model).

The next section describes the computational method. Following that, the calculation results for packing properties and objects are presented and discussed. Next, the computation results for the solubility of SWNTSs are given and examined. Later, the calculation results for the heat of solution of SWNTs are reported and analyzed. The last section summarizes the conclusions. Matrices will be denoted with capital letters. Assume that the data matrix X has dimensions ($n \times p$) where n denotes the number of observations and p the number of variables. A vector is always indicated in bold, *e.g.* $\mathbf{x}_i = (x_{i1}, \ldots, x_{ip})'$ stands for the i-th observation. Classical estimates are denoted by means of a tilde.

2 Computational Method

2.1 Solubility of Single-Wall Carbon Nanotubes

A new mechanism of solubility, which may explain the solubility of SWNTs of different sorts, is based on the possibility of formation in solution of clusters consisting of a number of SWNTs. This aggregation phenomenon changes the thermodynamic parameters of SWNTs in solutions, which displays, naturally, the phase equilibrium and changes the magnitude of solubility.

The thermodynamic approach to the description of the phenomenon of solubility of SWNTs, taking account cluster formation, is based on the bundlet model of clusters, which is valid under conditions when the characteristic number of SWNTs in the cluster is $n >> 1$.

Let us formulate the problem of determining the temperature dependence of solubility of SWNTs in terms of the possibility of forming clusters consisting of several parallel SWNTs. In accordance with the general principles of thermodynamics, in a saturated solution of SWNTs, they are equal the magnitudes of the chemical potential per SWNT for dissolved substance and for a crystal, which is in equilibrium with solution. This equality is valid not only for isolated SWNTs in a solution, but also for clusters consisting of several SWNTs. According to the bundlet model of clusters, the free energy of a cluster in a solution is made up of two parts, *viz.* the volume part, whose magnitude is proportional to the number of SWNTs n in the cluster, and the surface part proportional to $n^{1/2}$ [36–39]. Notice that the problem of growing a bundle of parallel cylinders in the 3D space can be reduced to that of growing a packet of circles on a plane. The model corresponds to the assumption that clusters consisting of $n >> 1$ particles have a cylindrical bundlet shape, and permits the Gibbs energy G_n for a cluster of size n to be represented as the sum

$$G_n = G_1 n - G_2 n^{1/2} \tag{1}$$

where the parameters G_1 and G_2 are responsible for the contribution to the Gibbs energy of molecules placed inside the volume and on the surface of a cluster, correspondingly. The chemical potential μ_n of a cluster of size n in a solution is expressed *via* the equation

$$\mu_n = G_n + T\ln C_n \tag{2}$$

where T is the temperature. Having regard to Equation (1), this results in

$$\mu_n = G_1 n - G_2 n^{1/2} + T\ln C_n \tag{3}$$

where the parameters G_1 and G_2 are expressed in temperature units.

In a saturated solution of SWNTs, the cluster-size distribution function is determined via the equilibrium condition linking the clusters of a specified size with a solid phase. This corresponds to the equality between the magnitudes of the chemical potential (per molecule) for molecules incorporated into clusters of any size and into crystal. This results in the following expression for the cluster-size distribution function in a saturated solution:

$$f(n) = g_n \exp\left(\frac{-An + Bn^{1/2}}{T}\right) \tag{4}$$

where the parameter A is the equilibrium difference between the energies of interaction of an SWNT with its surroundings in the solid phase and in the cluster volume; B is the similar difference for SWNTs located on the cluster surface; g_n is the statistical weight of a cluster of size n, which can depend, generally speaking, on both temperature and cluster size n. However, as is usually assumed, we will further neglect these dependences in comparison with the much stronger exponential dependence in Equation (4). The presented form (4) for the cluster-size distribution function is based on the structural features of SWNTs. In essence, an SWNT is a homogeneous surface structure that, unlike planar or elongated molecules, interacts with its surroundings almost irrespective of the orientation about its axis. The large number of similar elements of the SWNT surface makes it possible to represent the interaction energy of this molecule and the solvent molecules, having essentially smaller size, as the product of a specific surface interaction energy by surface area of the molecule. The feature of the SWNT structure may be further used in description of the interaction between clusters, made up of SWNTs, and the solvent. This is purely surface interaction and, because the interaction energy of SWNTs with one another, both in a cluster and in a solid is low in comparison with the binding energy of C atoms in an SWNT, one can assume that the specific surface energy of interaction of SWNTs with one another and with solvent molecules is not sensitive to the relative orientation of parallel SWNTs in a cluster.

Broadly speaking, the parameters A and B may have any sign. However, the normalization condition for the distribution function (4), having the obvious form

$$\sum_{n=1}^{\infty} f(n)n = C \tag{5}$$

causes the requirement $A > 0$. Here C is the solubility expressed in relative units. In view of the condition $n >> 1$, the normalization relationship (5) may be replaced by the integral

$$C = \bar{g}_n \int_{n=1}^{\infty} n \exp\left(\frac{-An + Bn^{1/2}}{T}\right) dn \tag{6}$$

Here $\bar{g}_n$ is the statistical weight of a cluster, averaged over the range of values of n that makes the major contribution to integral (6). This parameter includes the entropy factor, governs the absolute value of solubility and may depend on the type of solvent.

Expression (6) takes the following form

$$C = C_0 \int_{n=1}^{\infty} n \exp\left(\frac{-An + Bn^{1/2}}{T}\right) dn \tag{7}$$

which is applicable throughout the temperature range involved. Here C_0 is the SWNT molar fraction. In first approximation, parameters A, B and C_0 have been taken equal to those fitted for fullerene C_{60} in hexane, toluene and carbon disulphide (CS_2): $A = 320K$, $B = 970K$, $C_0 = 5 \cdot 10^{-8}$ (molar fraction) for $T > 260K$.

In order to improve the model, a correction has been introduced to take into account the effect of the different packing efficiencies in C_{60} and SWNTs

$$A' = \frac{\eta_{cyl}}{\eta_{sph}} A \quad \text{and} \quad B' = \frac{\eta_{cyl}}{\eta_{sph}} B \tag{8}$$

where $\eta_{cyl} = \pi/2(3)^{1/2}$ is the packing eficiency of cylinders (which is equal to that of circles on a plane), and $\eta_{sph} = \pi/3(2)^{1/2}$ is that of spheres (face-centred cubic, FCC).

The above-noted tendency of SWNTs in solution to form clusters is reflected in the magnitudes of parameters governing their properties in the conditions under consideration. Thus, the existence of the dependences of the cluster-size distribution function on the concentration and temperature of the solution leads to corresponding concentration and temperature dependences of thermodynamic and kinetic parameters, characterizing the behaviour of SWNTs in solutions. Specifically, in this Section it is shown that this phenomenon causes the concentration and temperature dependences of the heat of solution of SWNTs.

Using the approach developed above, let us state the form of the cluster-size distribution function in an unsaturated solution. In this case a solid phase is absent, so that the distribution function is determined via the equilibrium condition for clusters of various sorts in a solution. Using Equation (3) derived above for the chemical potential of a cluster in a solution, one can obtain the expression for the cluster-size distribution function in the unsaturated solution of SWNTs, depending on the solution concentration:

$$f_n(C) = \lambda^n \exp\left(\frac{-An + Bn^{1/2}}{T}\right) \tag{9}$$

Here the parameters A and B were determined above, and the parameter λ depending on the concentration of a solution is determined via the normalization condition

$$C = C_0 \int_{n=1}^{\infty} n\lambda^n \exp\left(\frac{-An + Bn^{1/2}}{T}\right) dn \tag{10}$$

The coefficient C_0 defines the absolute magnitude of the absolute concentration, and can be found by requiring (in specific conditions) the concentration determined *via* Equation (10) to be its saturated value, which is expressed via Equation (5). Saturation concentration has been taken as 0.0001mol·L^{-1}.

In terms of the above-described bundlet model of a cluster in a SWNT solution, the energy of formation of a cluster consisting of n SWNT units is determined by the following expression

$$E_n = n\left(An - Bn^{1/2}\right) \tag{11}$$

where the constants A and B were determined above. Using the expression for the cluster-size distribution function, one obtains the formula governing the thermal effect of solution of SWNTs per mole of dissolved substance:

$$H = \frac{\sum_{n=1}^{\infty} E_n f_n(C)}{\sum_{n=1}^{\infty} n f_n(C)} N_a = \frac{\sum_{n=1}^{\infty} n\left(An - Bn^{1/2}\right)\lambda^n \exp\left[\left(-An + Bn^{1/2}\right)/T\right]}{\sum_{n=1}^{\infty} n\lambda^n \exp\left[\left(-An + Bn^{1/2}\right)/T\right]} N_a \tag{12}$$

where the constant parameter λ is determined by the total concentration of the formed solution *via* the normalization condition (Equation 10).

2.2 Principal Component Analysis

Principal components analysis (PCA) is a dimension reduction technique [40]. From the original set of variables X_j, PCA constructs a new set of uncorrelated and orthogonal variables $\tilde{P}_j$ [41]. They are linear combinations of the mean-centred variables $\tilde{X}_j = X_j - \overline{X}_j$, and are often called the loadings or the principal components (PCs). It is well known that these loadings correspond with the eigenvectors of the sample covariance matrix $S = 1/(n-1)\sum_{i=1}^{n}(\mathbf{x}_i - \overline{\mathbf{x}})(\mathbf{x}_i - \overline{\mathbf{x}})$ of the data. For each loading vector $\tilde{P}_j$, the corresponding eigenvalue $\tilde{l}_j$ of S tells how much of the variability of the data is explained by $\tilde{P}_j$ through the relation $\tilde{l}_j = \text{Var}\left(\tilde{P}_j\right)$. Usually, these loading vectors are sorted in descending order of the eigenvalues. Hence, the first k PCs explain most of the variability of the data.

After selecting k, one can project the p-dimensional data points onto the subspace spanned by the k loading vectors, and compute their coordinates with respect to these $\tilde{P}_j$. This yields the scores

$$\tilde{\mathbf{t}}_i = \tilde{P}'\left(\mathbf{x}_i - \overline{\mathbf{x}}\right) \tag{13}$$

for each $i = 1, \ldots, n$, which have trivially zero mean. With respect to the original coordinate system, the projected data point is computed as the fitted value

$$\hat{\mathbf{x}}_i = \bar{\mathbf{x}} + \tilde{P}\tilde{\mathbf{t}}_i \qquad (14)$$

Notice that the $(p \times k)$ loading matrix $\tilde{P}$ contains the loadings column-wise. The $(k \times k)$ diagonal matrix $\tilde{L} = \left(\tilde{l}_j\right)_j$ will be used to denote the eigenvalues (in decreasing order).

To choose the appropriate number of loadings k, there exist many criteria. A popular graphical one is based on the scree plot, which exposes the eigenvalues in decreasing order. The index of the last component before the plot flattens is then selected. A more formal criterion considers the total variation that is explained by the first k loadings, and requires, *e.g.* that

$$\left(\sum_{j=1}^{k} \tilde{l}_j\right) \Bigg/ \left(\sum_{j=1}^{p} \tilde{l}_j\right) \geq 80\% \qquad (15)$$

PCs have two interesting properties: (1) they are extracted in decreasing order of importance; (2) each PC is orthogonal to each other.

2.3 Cluster Analysis

Cluster analysis (CA) [42] was applied to the packing properties. CA involved grouping the packing properties into clusters using hierarchical cluster analysis (HCA) [43]. The main purpose, here, was to form dissimilar clusters of the packing properties, to which the query packing objects would be compared to for determination of degree of similarity. We choose HCA for this similarity comparison over other more sophisticated similarity-determination methods because of its simplicity and speed. The results showed that its similarity-determination ability was sufficient for this application. HCA was performed using the set of packing properties $\{\text{closeness}, D, \eta\}$.

The aim of the CA was the recognition of groups of objects based on their similarity. It involves grouping a collection of objects into clusters (subsets), such that objects within each cluster are more closely related to one another than objects in different clusters. In this study HCA was used. In HCA the objects are clustered into subgroups in a series of partitions. The basic process of HCA starts by assigning each object as a cluster, so that N objects will form N clusters with each cluster containing just one item. Then the similarities between the clusters are determined by the distances (similarities) between the items they contain. The next step is to find the most similar (closest) pair of clusters and merge them into a new cluster, so that now the number of clusters becomes $N - 1$. Then the distances (similarities) between the newly formed cluster and each of the old clusters are computed. The previous two steps are iterated until all items are clustered into a single cluster containing N objects. Within HCA there are two subdivisions: agglomerative and divisive methods. The algorithm just described is an agglomerative method and it is the method implemented in this study. The divisive method works in the opposite way. It separates N objects into finer groups. Every object is initially in a huge cluster. Then the cluster is divided continuously until the desired number of clusters is formed. In brief, agglomerative method is a *bottom-up* clustering method whereas divisive method uses a *top-down* approach.

In HCA the distances (similarities) computation (distance calculation between clusters) can be done in different ways (*cf.* Figure 1). This cluster fusion step is often called the *linkage* method, which is what distinguishes single- from average- and complete-linkage CA. In *single-linkage CA*, the shortest distance from any object in one cluster to any object in the other cluster is considered as the distance between one cluster and another cluster. In *average-linkage CA*, the distance between two clusters is the average distance from any object in one cluster to any object in the other cluster. *Complete-linkage CA* considers the longest distance from any object in one cluster to any object in the other cluster. In this study, single- and complete-linkage CA were used.

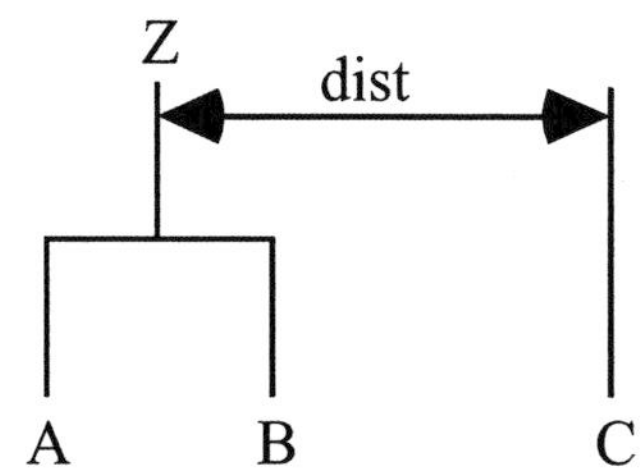

Figure 1. Distance between clusters Z and C.

Single linkage is one of the simplest agglomerative HCA methods. The distance between clusters is given by the value of the shortest link between the clusters. The distance $D(x,y)$ is computed as

$$D(x, y) = \mathrm{Min}\left[d(i, j)\right] \tag{16}$$

where object i is in cluster x and object j is in cluster y.

At each stage of HCA the cluster x and y for which $D(x,y)$ is minimum are merged. In average-linkage method the distance $D(x,y)$ is computed as

$$D(x, y) = \frac{T_{xy}}{N_x N_y} \tag{17}$$

where T_{xy} is the sum of all pairwise distances between cluster x and y. N_x and N_y are the sizes of the clusters x and y, respectively

At each stage of HCA the cluster x and y for which $D(x,y)$ is minimum are merged. Briefly, the distance between two clusters is defined as the average of distances between all pairs of objects, where each pair is made up of one object from each group.

Various types of HCA methods differ not only in the linkage method used to merge objects, but also in the distance metric used to calculate the distance matrix between objects. Frequently used similarity metrics include Euclidean distance, standardized Euclidean distance, city-block (Metropolis) distance, Minkowski distance and Mahalanobis distance. In this study, standardized Euclidean distance was selected as it provided the best CA. Standardized Euclidean distance was calculated as by the following equation,

$$d_{xy} = \left[(T_x - T_y)D^{-1}(T_x - T_y)\right]^{1/2} \quad (18)$$

where **T** is an $m\times n$ matrix, d_{xy} is the distance between the vector T_x and T_y. **D** is the diagonal matrix with diagonal elements given by the variance of the variable T_i over the m objects.

City-block (Metropolis) distance is simply the summed difference across dimensions and was calculated as

$$d_{xy} = \sum_{i=1}^{n} \left|T_{xi} - T_{yi}\right| \quad (19)$$

With the formation of clusters, the similarity was calculated between the query properties and the clusters. The distance measured used to determine this similarity was the distance from the query property to the centroid of the cluster in the feature space.

There are many reasons why one might want to cluster a database of molecular structures [44–47].

A program has been written using the IMSL [48] subroutine CLINK to carry out HCA, based upon either a *distance* or a *similarity* (*e.g.*, correlation) *matrix*.

3 Calculation Results for Packing Properties

Table 1 summarizes the packing parameters *closeness*, packing dimension D, efficiency η and packing-efficiency correction factor of equal objects, for clusters and other systems of atoms with short-range interaction.

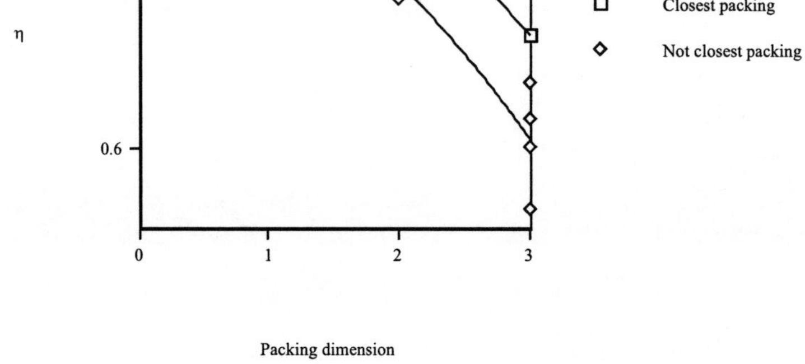

Figure 2. Variation of the packing efficiency η *vs.* packing dimension D for closest and not closest packings.

Table 1. Closeness, packing dimensions D, efficiencies η and packing-efficiency correction factors for equal objects.

Objects	Closeness	D	Packing efficiency
Sphere simple cubic (SC)	Not closest	3	$\frac{\pi}{6} \approx 0.52360$
Sphere random *loose* (RL)	Not closest	3	0.601±0.005
Sphere random close (RC)	Not closest	3	0.6366±0.0005
Sphere body-centred cubic (BCC)	Not closest	3	$\frac{\pi\sqrt{3}}{8} \approx 0.68017$
Sphere (FCC alias cubic closest packing, CCP or hexagonal closest packing, HCP)	Closest	3	$\frac{\pi}{3\sqrt{2}} \approx 0.74048$
Cylinder in space as square packing (SP) of circles on a plane	Not closest	2	$\frac{\pi}{4} \approx 0.78540$
Cone (solid angle 1 stereoradian)	Not closest	2	$\sqrt{1-\frac{1}{\pi}} \approx 0.82565$
Cylinder (as hexagonal packing of circles on a plane)	Closest	2	$\frac{\pi}{2\sqrt{3}} \approx 0.90690$
Cube	Closest	0	1.0
η_{con}/η_{sph}	–	0.66667	$\frac{3\sqrt{2\left(1-\frac{1}{\pi}\right)}}{\pi} \approx 1.11501$
η_{cyl}/η_{sph}	–	0.66667	$\sqrt{\frac{3}{2}} \approx 1.22474$

For closest and not closest packings, the variations of the packing efficiency η with the packing dimension D (*cf.* Figure 2) show that from $D = 2$ to 3 $\eta_{\text{not closest}}$ drops quicker than η_{closest}.

The changes can be represented, respectively, by

$$\eta_{\text{closest}} = 1.00 + 0.0334D - 0.0400D^2 \tag{20}$$

$$\eta_{\text{not closest}} = 1.00 - 0.0319D - 0.0326D^2 \quad r = 0.954 \tag{21}$$

Table 2 reports the packing efficiencies, packing-efficiency correction factors and the numerical values of the parameters A, B and C_0 determining the energy of interaction of molecules.

Table 2. Packing efficiencies η, correction factors and parameters determining the energy of interaction of molecules.[a]

Molecule	η	η-correction factor	A (K)	B (K)
C_{60}-face-centred cubic[b], SWNH[c], SWNT[d]	0.74048	1.0	320	970
SWNH[c] η-correction	0.82565	1.11501	357	1082
SWNT[d] η-correction	0.90690	1.22474	392	1188

[a] $C_0 = 5\cdot10^{-8}$ (molar fraction).
[b] For $T > 260$K.
[c] SWNH: single-wall carbon nanohorn.
[d] SWNT: single-wall carbon nanotube.

Packing property *closeness* (Table 1) has been categorized as 0 and 1 in arbitrary units for *not closest* and *closest*, respectively. The upper triangle of the 3x3 symmetrical correlation matrix **R** calculated for the packing properties {closeness,D,η} results:

$$\mathbf{R} = \begin{pmatrix} 1.000 & -0.500 & 0.681 \\ & 1.000 & -0.864 \\ & & 1.000 \end{pmatrix}$$

High absolute correlation is observed between D and η. The absolute correlation between D and η is double than that between *closeness* and D. The absolute correlation increases in the order $|R_{\text{closeness},D}| < |R_{\text{closeness},\eta}| < |R_{D\eta}|$.

Both single- and complete-linkage HCAs allow building the *dendrogram* (binary tree) for the packing properties {closeness,D,η} (*cf.* Figure 3) [49]. Both HCAs perform a binary taxonomy of {closeness,D,η} that separates first property *closeness* from D and η; finally, D is set apart from η. The earliest separation of property *closeness* is in agreement with the high value of $|R_{D\eta}|$.

From both HCAs the radial tree for the packing properties {closeness,D,η} is represented in Figure 4. The packing properties D and η are shown more similar than property *closeness* in agreement with the dendrogram.

SplitsTree is a program for analyzing CA data [50]. Based on the method of *split decomposition*, SplitsTree takes as input a *distance matrix* or a set of CA data, and produces as output a graph that represents the relationships between the taxa. For ideal data this graph is a tree, whereas less ideal data will give rise to a tree-like network, which can be interpreted as possible evidence for different and conflicting data. Further, as split decomposition does not attempt to force data onto a tree, it can provide a good indication of how *tree*-like given data are. The splits graph for the packing properties {closeness,D,η} (*cf.* Figure 5) suggests a conflicting relationship between D and η, due to the D–η interdependence. Therefore, the splits graph indicates a spurious relationship resulting from base-composition effects. This increases the risk of co-linearity [51] given the D–η relationship.

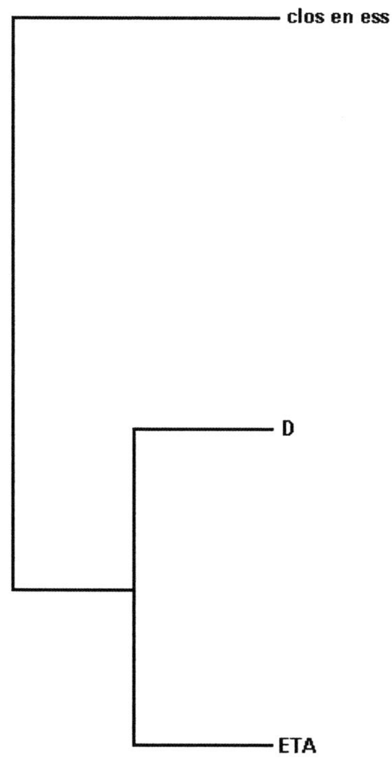

Figure 3. Dendrogram for the packing properties *closeness*, *D* and η (ETA) of the different packing objects.

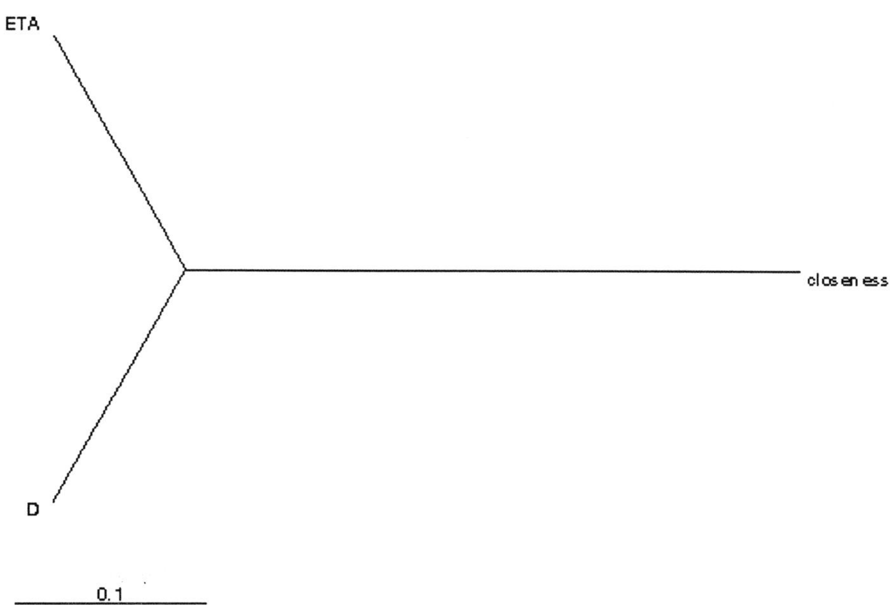

Figure 4. Radial tree for the packing properties *closeness*, *D* and η (ETA) of the different packing objects.

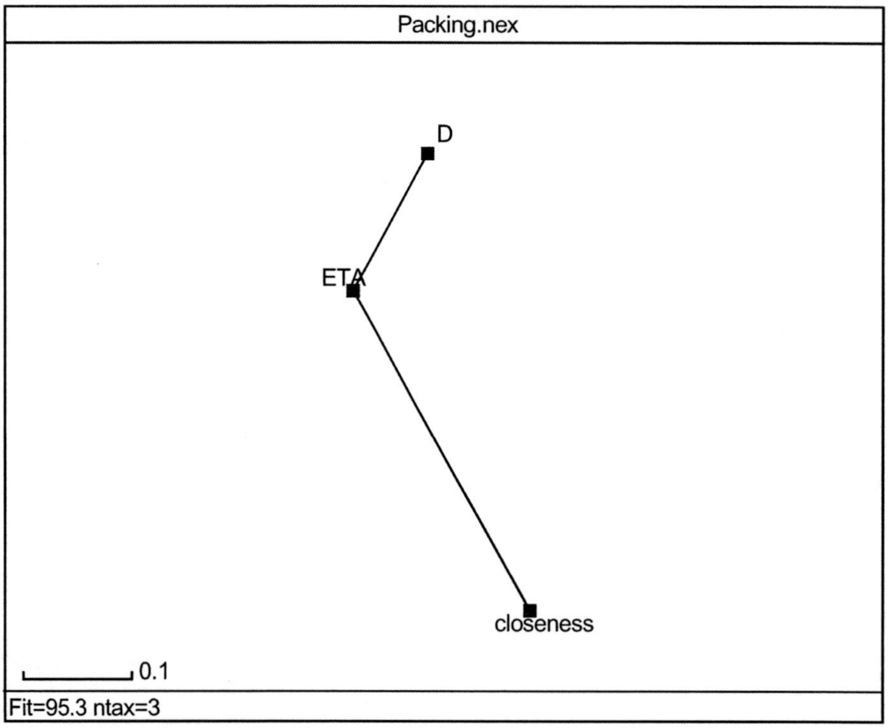

Figure 5. Splits graph for the packing properties *closeness*, *D* and η (ETA) of the different packing objects.

The PCA for the packing properties {closeness,D,η} results in three factors, which are linear combinations of *closeness*, *D* and η. The importance of the PCA factors F_1–F_3 for the packing properties of equal objects is collected in Table 3. In particular, the use of only the first factor F_1 explains 79% of the variance (21% error); the combined use of the first two factors $F_{1\text{–}2}$ explains 97% of the variance (3% error).

The PCA factor loadings are shown in Table 4.

Table 3. Importance of the principal component analysis factors.

Factor	Eigenvalue	Percentage of variance	Cumulative percentage of variance
F_1	2.37397342	79.13	79.13
F_2	0.52262988	17.42	96.55
F_3	0.10339670	3.45	100.00

Table 4. Principal component analysis loadings for the structural parameters.[a]

	PCA factor loadings		
Property	F_1	F_2	F_3
closeness	0.521	**0.817**	-0.245
D	-0.582	0.551	0.598
□	0.624	-0.169	**0.763**

[a] Loadings greater than 0.7 are boldfaced.

The PCA factors F_{1-3} profile for the packing properties is summarized in Table 5. In particular, for both F_1 and F_3 variable η presents the greatest weight in the profile; however, F_1 cannot be reduced to two variables $\{D,\eta\}$ without a 27% error. For F_2 variable *closeness* shows the greatest weight; F_2 can be reduced to two variables {closeness,D} with a 3% error. In some way, factors F_1 and F_3 could be considered as linear combinations of D and η (with errors of 27% and 6%, respectively). Furthermore, factor F_2 can be expressed as a linear combination of *closeness* and D with an error of only 3%.

Table 5. Profile of the principal component analysis factors.[a]

Factor	Percentage of *closeness*	Percentage of D	Percentage of η
F_1	27.15	33.89	38.96
F_2	**66.83**	30.32	2.85
F_3	6.02	35.79	**58.19**

[a] Percentages greater than 50% are boldfaced.

In the principal component analysis (PCA) F_1 *vs.* F_2 plot for the packing properties, four classes of equal objects are clearly distinguished, *viz.* class 1 with two units ($F_1 < F_2$, Figure 6 *left*), class 2 (two units, $0 \approx F_1 > F_2$, *bottom*), class 3 (three units, $F_1 > F_2$, *right*), and class 4 (two units, $F_1 < F_2 \approx 0$, *middle left*). In general the packings of equal spheres, *viz.* sphere SC, sphere RL, sphere RC, sphere BCC, and *sphere* (FCC) belong to two classes with the only exception of the packing *sphere* (FCC). The closest packings, *viz.* sphere, cylinder, and cube belong to the same class. The packings with the highest efficiencies η, *viz.* cylinder, and cube belong to the same class. The packings with the lowest efficiencies η, *viz.* sphere SC, sphere RL, sphere RC, and sphere BCC belong to two classes.

On the other hand, instead of nine objects in the space of three packing properties (Table 1), consider three properties in the space of nine objects. A table with three rows and nine columns was built and the similarity of the packing objects was compared. From both HCAs, the dendrogram for the packing objects matching to the properties {closeness,D,η} provides a binary taxonomy of the objects (Table 1), which separates first both units in class 1 (sphere SC and sphere BCC, Figure 8 *top*), then class 2 (two units, *viz.* cylinder SP, and cone, *middle top*), class 3 (three units, *viz.* cube, sphere, and cylinder, *middle bottom*), and finally class 4 (two unis, *viz.* sphere RL and sphere RC, *bottom*). The classes correspond to those obtained by PCA (Figures 6 and 7).

From both HCAs, the radial tree for the packing objects relating to {closeness,D,η} separates first both units in class 1 (sphere SC and sphere BCC, Figure 9 *right*), then class 2 (two units, *viz.* cylinder SP, and cone, *bottom*), class 3 (three units, *viz.* cube, sphere, and cylinder, *left*), and finally class 4 (two unis, *viz.* sphere RL and sphere RC, *top*). The classes correspond to the PCA and dendrogram (Figures 6, 7 and 8).

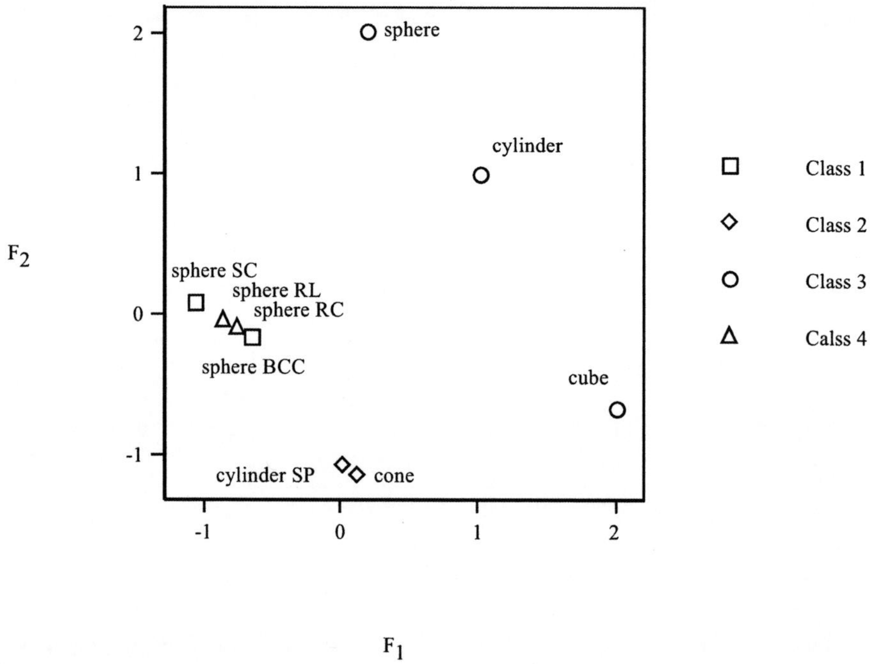

Figure 6. Principal component analysis F_1 *vs.* F_2 plot for the different packing properties. The PCA F_1 *vs.* F_3 plot for the packing properties separates somewhat better class 1 from class 4.

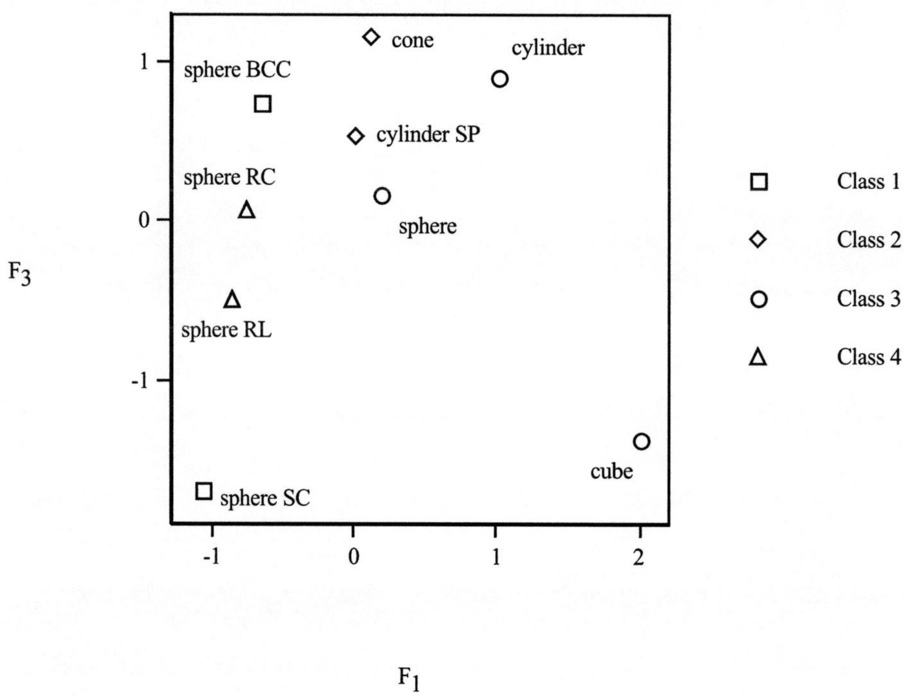

Figure 7. Principal component analysis F_1 *vs.* F_3 plot for the different packing properties.

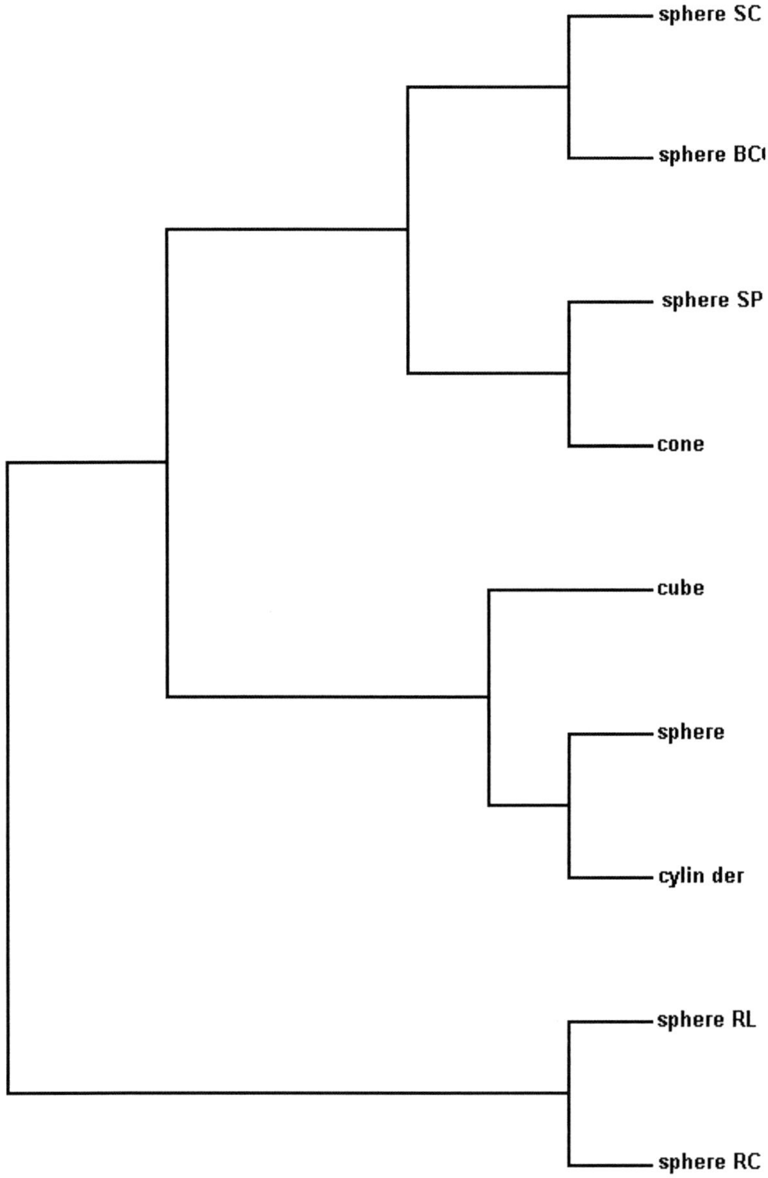

Figure 8. Dendrogram for the packing objects.

Figure 10 displays the splits graph for the packing objects. The packing objects sphere (RL) and sphere (RC) appear superimposed. The splits graph reveals conflicting relationships between the sphere SC–sphere BCC, sphere SC–cube and cylinder–cube pairs, due to sphere SC–sphere BCC, sphere SC–cube and cylinder–cube interdependences. Therefore, the splits graph indicates spurious relationships resulting from base-composition effects. These increase the risk of co-linearity given the sphere SC–sphere BCC, sphere SC–cube and cylinder–cube relationships.

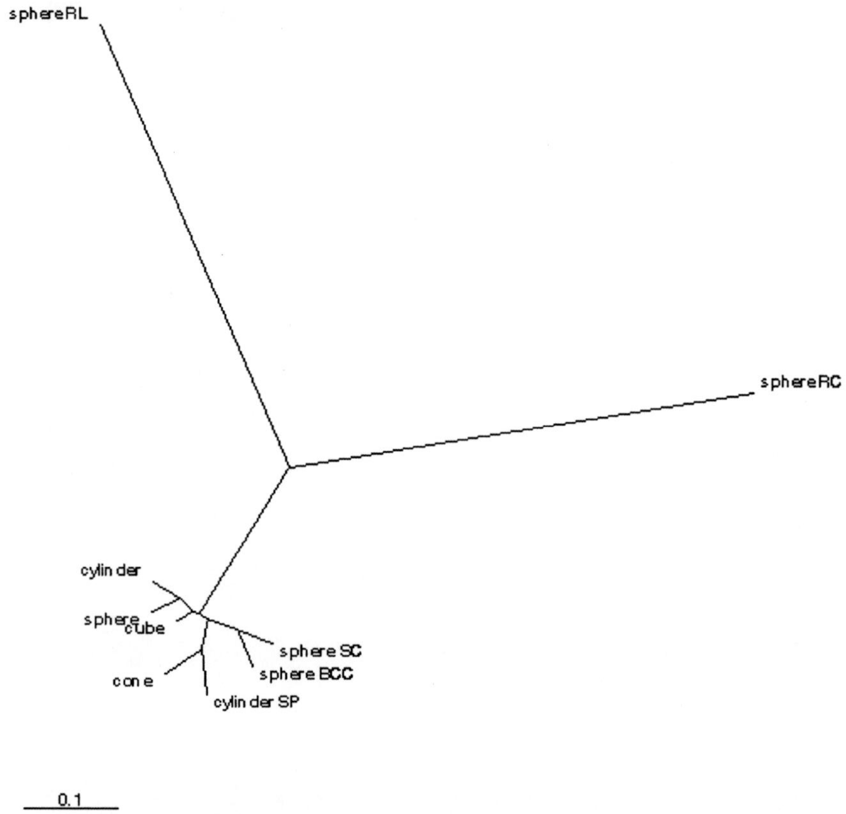

Figure 9. Radial tree for the packing objects.

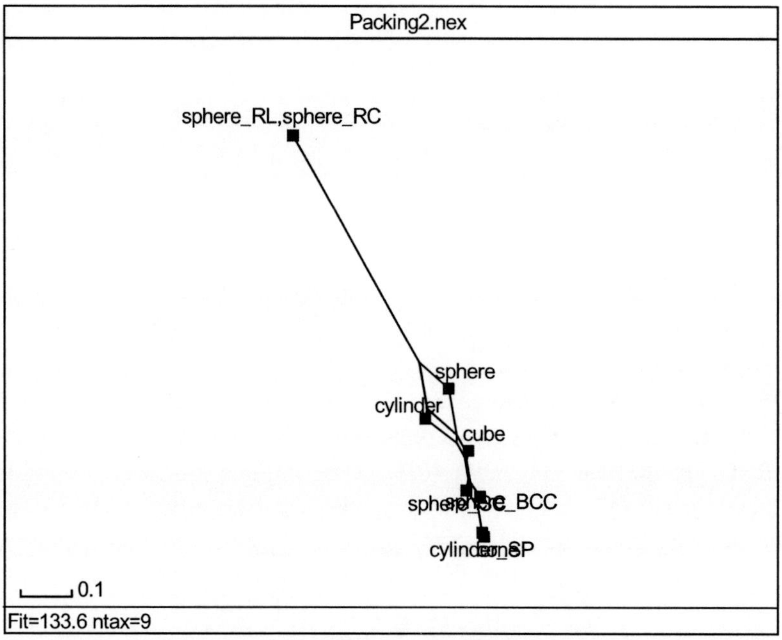

Figure 10. Splits graph for the packing objects.

4 Calculation Results for Solubility of Single-Wall Carbon Nanotubes

Figure 11 illustrates the equilibrium difference between the Gibbs free energies of interaction of an SWNT with its surroundings, in the solid phase and in the cluster volume, or on the cluster surface. On going from C_{60} (droplet model) to SWNT (bundlet model) the minimum is less marked (55% of droplet model). This causes a lesser number of units in SWNT ($n_{minimum} \approx 2$) than in C_{60} clusters (≈8). Moreover, the abscissa is also shorter in SWNT (≈9) than in C_{60} clusters (≈28). In turn, when the packing-efficiency correction (Equation 8) is included, the C_{60}–SWNT shortening decreases (68% of droplet model) while keeping $n_{minimum} \approx 2$ and $n_{abscissa} \approx 9$.

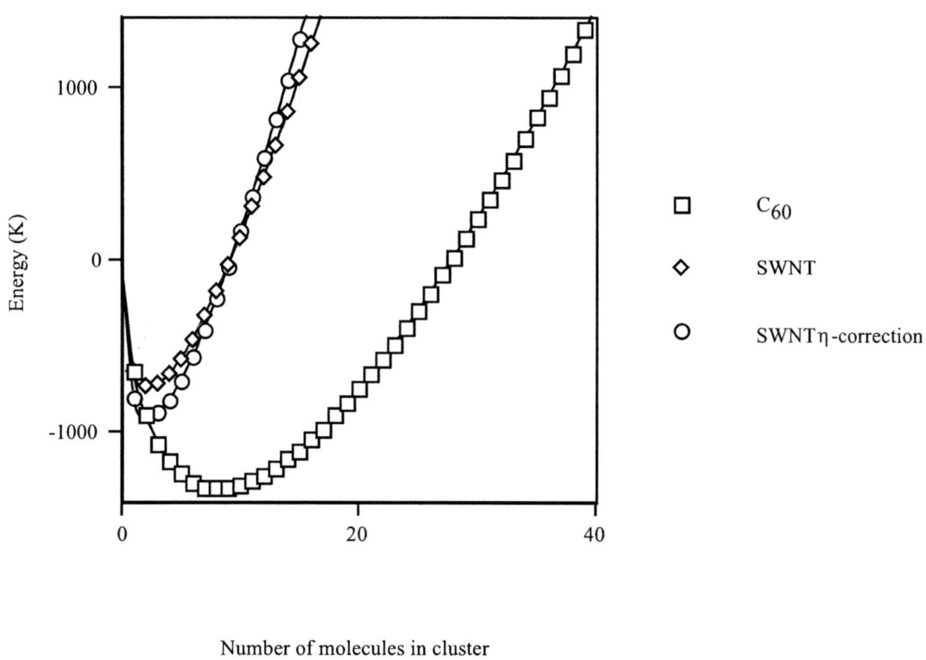

Figure 11. Energy of interaction of an SWNT with its surroundings in the cluster volume or surface.

The temperature dependence of the solubility of SWNT is shown in Figure 12. The solubility decreases with temperature, because solubility is due to cluster formation. At T ≈ 260K the C_{60} crystal presents an orientation disorder phase transition from face-centred cubic (FCC) lattice characterized by close packing to simple cubic (SC, Table 1) lattice. The reduction is less marked for SWNT, in agreement with the lesser number of units in SWNT clusters (Figure 11). In particular at T = 260K on going from C_{60} (droplet model) to SWNT (bundlet model), the solubility drops to 1.6% of droplet (Figure 12). In turn, when the packing-efficiency correction is included (Equation 8) the shortening decreases (2.6% of droplet model).

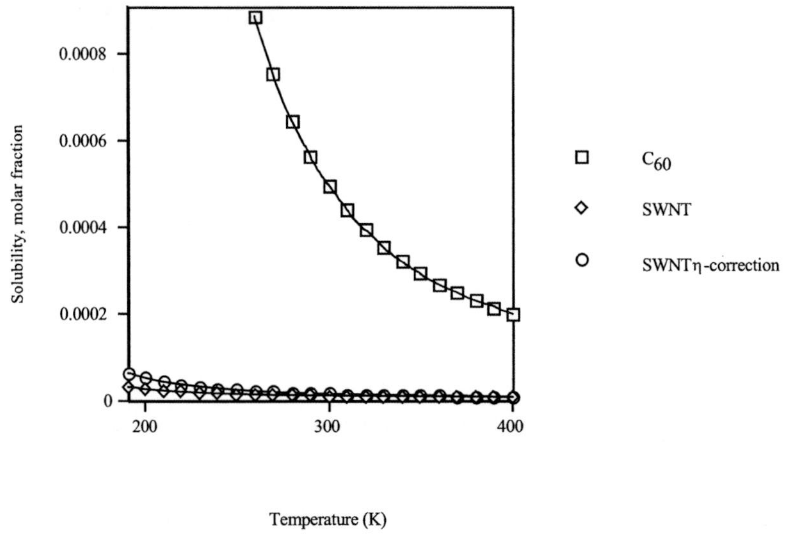

Figure 12. Temperature dependence of solubility of C_{60} (droplet) and SWNT (bundlet model without and with η-correction).

The cluster distribution function by size in a solution of SWNT in CS_2, calculated for saturation concentration at solvent temperature T = 298.15K is displayed in Figure 13. On going from C_{60} (droplet model) to SWNT (bundlet model), the maximum cluster size decreases from n = 8 to 2 and the distribution is rather narrowed, in agreement with the lesser number of units in SWNT clusters (Figures 11 and 12). On the other hand when the packing-efficiency correction (Equation 8) is included, the distribution is somewhat narrowed while keeping $n_{maximum} = 2$ (Figure 13).

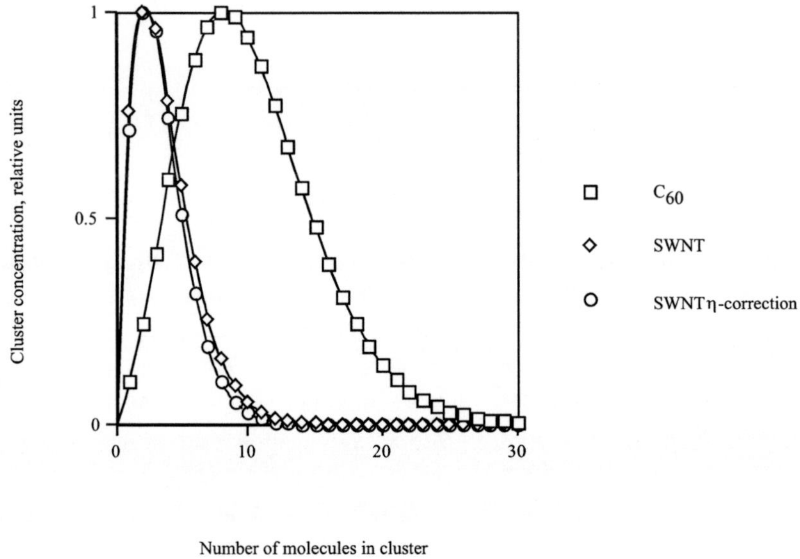

Figure 13. Cluster distribution by size saturated in CS_2 at T = 298.15K of C_{60} (droplet) and SWNT (bundlet/η-correction).

5 Calculation Results for the Heat of Solution of Single-Wall Carbon Nanotubes

The concentration dependence of the heat of solution in toluene, benzene and CS_2, calculated at solvent temperature T = 298.15K, is presented in Figure 14. For C_{60} (droplet model), on going from a concentration of solution less than 0.1% of saturated (containing only isolated fullerene molecules) to that with concentration 15% (containing clusters of average size 7), the heat of solution decreases by 73%. In turn for SWNT (bundlet model), the heat of solution increases by 98% in the same interval. However, when the packing-efficiency correction (Equation 8) is included, the increment in the heat of solution is reduced to 54% in the same interval. Apparently, the discrepancy between various experimental data on the heat of solution of fullerenes and SWNTs may be ascribed to such a sharp concentration dependence of the heat of solution of fullerenes and SWNTs.

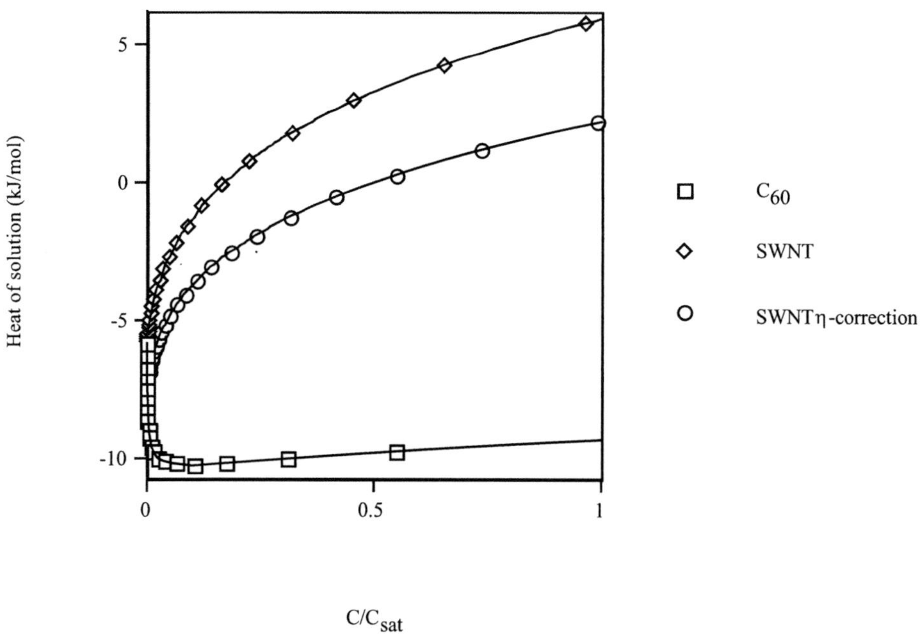

Figure 14. Concentration dependence of the heat of solution in toluene, benzene and CS_2 at T = 298.15K of C_{60} and SWNT.

The temperature dependence of the heat of solution in toluene, benzene and CS_2, calculated for saturation concentration, is exhibited in Figure 15. At T ≈ 260K the C_{60} crystal shows a transition from FCC to SC. For C_{60} (droplet model), on going from T = 260K to T = 400K, the heat of solution increases ca. $3kJ \cdot mol^{-1}$. For SWNT (bundlet model), the heat of solution augments $11kJ \cdot mol^{-1}$ in the same interval. In turn, when the packing-efficiency correction (Equation 8) is included, the increase in the heat of solution is reduced to $10kJ \cdot mol^{-1}$ in the same interval.

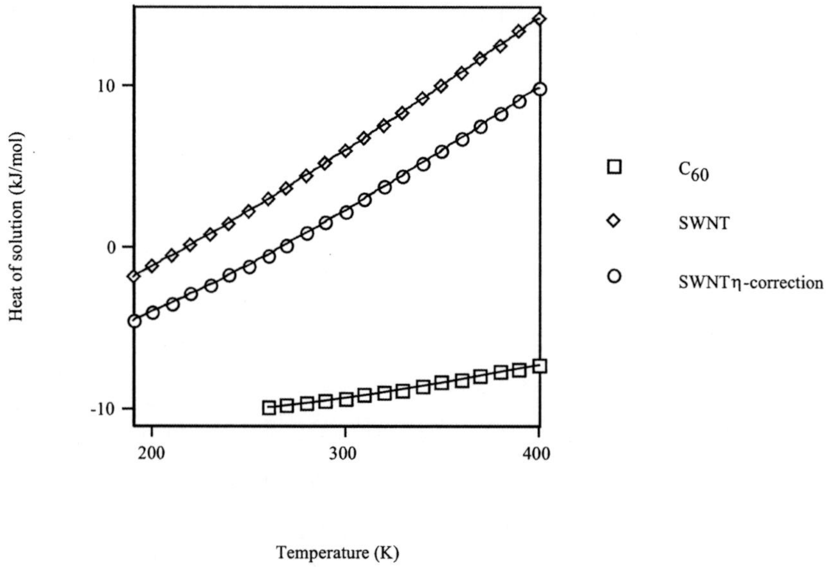

Figure 15. Temperature dependence of heat of solution in toluene, benzene and CS_2 for saturation of C_{60}/SWNT.

As follows from the given data, cluster formation in fullerene and SWNT solutions has a considerable impact on the thermal effect in a solution, and results in concentration and temperature dependences of the heat of solution. This phenomenon should be taken into consideration in measuring the thermal effect of solution of fullerenes, SWNTs and other substances forming aggregates in solutions. On the other hand, the existence of a concentration dependence of the heat of solution of a substance may be used as evidence of the cluster origin of the solubility of this substance.

6 Conclusion

The following conclusions can be made from this study.

1. Close packings are the tightest way to pack spheres. Atoms and fullerenes, being nothing but tiny spheres often arrange in this way. Furthermore, it is also possible to deduce atomic structures of metal alloys, salts and oxides by fitting the voids of close-packed spheres.
2. Several criteria have been selected to reduce the analysis to a manageable quantity of packing properties and objects from the enormous set of equal objects. The former ones refer to three properties, *viz.* packing closeness, packing dimension, and packing efficiency.
3. The area of clustering is notoriously difficult; *e.g.*, although oranges and apples seem to have significant differences, they are both fruit. Is a pomegranate more like an apple or is it more like an orange? When the clustering problem is poorly specified, or the variation within each cluster is greater than that between different clusters, meaningful clustering often becomes almost impossible. Progression in the

development of new methods is hampered by the lack of *gold standards*, against which to judge the quality of any clustering exercise. An understanding of both the chemistry and the computational methods is essential for tackling the associated *data mining* tasks, without being distracted by the abundant fool's gold. If a small number of clusters of data are easy to fit, the predictive ability of the model could be guaranteed only if the deviations inside the clusters do not diverge.

4. For the dataset employed in this study, we found that a model can reliably predict a packing object's property if the object is sufficiently similar in structure to the group of objects used to generate that model. Therefore, the two-step approach involving object clustering and property prediction provided an efficient way to assess the reliability of a model's prediction. In conclusion, the new, non-computationally intensive and easily understood approach designed in this study not only solved the problem, but also should have practical applications in the modelling field [52]. Future work may include examining the approach with more datasets with different properties.
5. The cluster analysis shows greater similarity for the packing properties D and η than when comparing with *closeness*. Split decomposition suggests a spurious relationship resulting from base-composition effects.
6. Both single- and complete-linkage cluster analyses of the packing properties of equal objects, clusters and other systems of atoms with short-range interaction, allow classifying these properties and agree. Principal components analysis of the packing properties and cluster analysis of the packing objects permit classifying them and agree. The packing objects have been grouped into three classes.
7. For a cluster nature of the SWNT solubility to be completely established, direct experimental exploration is necessary. As a possible scheme for such experiments, the measurements of infrared absorption spectra of an SWNT solution, involving concentrations at various temperatures and a constant optical path length, can be conceived. The existence of such a dependency will indicate the presence of SWNT clusters in solution. Another way is based on applying Raoult's law. According to this law, the saturation vapour pressure of a solvent above a solution differs from that above a pure solvent by a value being proportional to the concentration of solute particles. The measurement of the flow of the solvent vapour, being determined by the difference of the pressures, will enable obtaining the dependence of the solute particle concentration on the concentration of the solute. Provided this dependence is nonlinear, it will form a basis to draw a quantitative conclusion concerning the existence of the SWNT clusters in solution.
8. Along with solubility, the cluster formation also must affect the transport properties of SWNTs in solution, particularly diffusion. It can be shown that a characteristic size of clusters decreases as the SWNT concentration drops below the saturation value. Since the mobility of a cluster increases with its decreasing size, it leads to the dependence of an effective diffusion coefficient on concentration. This can serve as a basis for the method of diffusion separation of higher SWNTs, being usually of a tiny concentration in solution.
9. Fullerenes–SWNTs are unique objects of chemical physics, whose behaviour in many physical situations is characterized by remarkable peculiarities. Peculiarities in the behaviour of fullerenes–SWNTs in solutions show up first in that fullerenes–

SWNTs represent the only soluble form of carbon. This is primary related to the originality in the molecular structure of fullerenes–SWNTs. The fullerene molecule (SWNT) is a virtually uniform closed spherical or spheroidal (cylindrical) surface having no sharp ridges or dents. Both structures give rise to the relatively weak interaction between the neighbouring molecules in a fullerite crystal, and promote effective interaction of the molecules with those of a solvent.

10. Another interesting peculiarity of the behaviour of fullerenes–SWNTs in solutions is related to their tendency to form clusters consisting of a number of fullerene molecules–SWNTs. The energy of interaction of a fullerene molecule–SWNT with solvent molecules is proportional to the surface of the former molecule, and roughly independent of the relative orientation of solvent molecules.
11. The performed analysis of experiments concerning the behaviour of fullerenes–SWNTs in solutions shows that, over a wide range of conditions, fullerenes in solutions exist in the form of aggregates containing from ten up to tens of thousands of particles. Therefore, fullerenes in solutions represent a unique case of a cluster substance, the thermodynamically equilibrium state of which corresponds to the majority of the matter in the cluster form. Such a unique peculiarity, attributed to details in the interaction of fullerene molecules with one another and with solvent molecules, can be realized only in solutions, whereas the cluster substance in the gas phase can be produced only in highly nonequilibrium conditions [53], *e.g.*, in the conditions of a supersonic nozzle. However, the size of clusters of SWNTs is much smaller. This peculiarity makes fullerenes–SWNTs in solutions interesting objects from the viewpoint of the investigation of the properties of a cluster substance.
12. The aggregation of fullerenes–SWNTs is reflected in various physical and chemical properties of fullerene solutions. Thus, this phenomenon causes the extraordinary (decreasing) temperature dependence of solubility.
13. All these phenomena were observed in experiments, and have a unified explanation in the framework of the droplet–bundlet models of a cluster, in accordance with which the free energy of a fullerene molecule (SWNT) involved in a cluster is combined from two components: a volume one, which is proportional to the number of molecules n in a cluster, and a surface one, which is proportional to $n^{2/3}$ ($n^{1/2}$).
14. At present, experimental SWNT properties depend on the sample, since some samples: (1) contain fullerenes, (2) consist of diameters of metallic–semimetallic or semiconductor SWNTs, (3) show polydispersity between short and large SWNTs; (4) solubility of SWNTs differ for different diameters; (5) SWNTs thinner than (5,5) are scarce. The tendency of fullerenes–SWNTs to form clusters in solutions must be taken into consideration when further developing the technology of enrichment, separation and purification of fullerenes–SWNTs.
15. From purely geometrical differences, the droplet and bundlet models of clusters predict different behaviours due, in part, to the fact that the packing efficiency of cylinders is greater than that of spheres. As the sensitivity to packing efficiency is smaller in cylinders, packing-efficiency corrections are expected smaller for SWNTs.

It has not escaped our notice a droplet model for clusters of conical single-wall carbon nanohorns (SWNHs), following modified Equations (1’)–(12’).

$$G_n = G_1 n - G_2 n^{2/3} \tag{1'}$$

$$\mu_n = G_1 n - G_2 n^{2/3} + T \ln C_n \tag{3'}$$

$$f(n) = g_n \exp\left(\frac{-An + Bn^{2/3}}{T}\right) \tag{4'}$$

$$C = \bar{g}_n \int_{n=1}^{\infty} n \exp\left(\frac{-An + Bn^{2/3}}{T}\right) dn \tag{6'}$$

$$C = C_0 \int_{n=1}^{\infty} n \exp\left(\frac{-An + Bn^{2/3}}{T}\right) dn \tag{7'}$$

$$A' = \frac{\eta_{\text{con}}}{\eta_{\text{sph}}} A \quad \text{and} \quad B' = \frac{\eta_{\text{con}}}{\eta_{\text{sph}}} B \tag{8'}$$

where η_{con} is the packing efficiency of cones, *e.g.*, for a solid angle of one stereoradian, $\eta_{\text{con}} = (1-1/\pi)^{1/2}$. As $\eta_{\text{sph}} < \eta_{\text{con}} < \eta_{\text{cyl}}$ (Table 1), the behaviour of conical SWNHs is expected to be intermediate between those of spherical fullerenes and cylindrical SWNTs.

$$f_n(C) = \lambda^n \exp\left(\frac{-An + Bn^{2/3}}{T}\right) \tag{9'}$$

$$C = C_0 \int_{n=1}^{\infty} n\lambda^n \exp\left(\frac{-An + Bn^{2/3}}{T}\right) dn \tag{10'}$$

$$E_n = n\left(An - Bn^{2/3}\right) \tag{11'}$$

$$H = \frac{\sum_{n=1}^{\infty} E_n f_n(C)}{\sum_{n=1}^{\infty} n f_n(C)} N_a = \frac{\sum_{n=1}^{\infty} n\left(An - Bn^{2/3}\right)\lambda^n \exp\left[\left(-An + Bn^{2/3}\right)/T\right]}{\sum_{n=1}^{\infty} n\lambda^n \exp\left[\left(-An + Bn^{2/3}\right)/T\right]} N_a \tag{12'}$$

Acknowledgements

The authors acknowledge financial support from the Spanish MEC DGI (Project No. CTQ2004-07768-C02-01/BQU) and Generalitat Valenciana (DGEUI INF01-051 and INFRA03-047, and OCYT GRUPOS03-173).

References

[1] Ercolessi, F. A Molecular Dynamics Primer; *International School for Advanced Studies: Trieste*, 1997 pp 7-7.

[2] Ruoff, R. S.; Malhotra, R.; Huestis, D. L.; Tse, D. S.; Lorents, D. C. *Nature* (London) 1993, 362, 140-141.

[3] Blau, W. J.; Byrne, H. J.; Cardin, D. J.; Dennis, T. J.; Hare, J. P.; Kroto, H. W.; Taylor, R.; Walton, D. R. M. *Phys. Rev. Lett.* 1991, 67, 1423-1425.

[4] Sun, Y.-P.; Bunker, C. E. *Nature* (London) 1993, 365, 398-398.

[5] Ghosh, H. N.; Sapre, A. V.; Mittal, J. P. *J. Phys. Chem.* 1996, 100, 9439-9443.

[6] Ying, Q.; Marecek, J.; Chu, B. *Chem. Phys. Lett.* 1994, 219, 214-218.

[7] Ying, Q.; Marecek, J.; Chu, B. *J. Chem. Phys.* 1994, 101, 2665-2672.

[8] Suzuki, K. *In Molecular Nanostructures*; Kuzmany, H.; Fink, J.; Mehring, M.; Roth, S.; Eds.; World Scientific: Singapore, 1998.

[9] Painter, P. C.; Opaprakasit, P.; Scaroni, *A. Energy Fuels 2000*, 14, 1115-1118.

[10] Scrivens, W. A.; Tour, J. M.; Creek, K.E.; Pirisi, L. *J. Am. Chem. Soc.* 1994, 116, 4517-4518.

[11] Fortner, J. D.; Lyon, D. Y.; Sayes, C. M.; Boyd, A. M.; Falkner, J. C.; Hotze, E. M.; Alemany, L. B.; Tao, Y. J.; Guo, W.; Ausman, K. D.; Colvin, V. L.; Hughes, J.B. *Environ. Sci. Technol.* 2005, 39, 4307-4316.

[12] Andrievsky, G. V.; Kosevich, M. V.; Vovk, O. M.; Shelkovsky, V. S.; Vashchenko, L. A. *J. Chem Soc. Chem. Commun.* 1995, 1281-1282.

[13] Deguchi, S.; Alargova, R. G.; Tsujii, K. *Langmuir* 2001, 17, 6013-6017.

[14] Andrievsky, G. V.; Klochkov, V. K.; Bordyuh, A. B.; Dovbeshko, G. I. *Chem Phys. Lett.* 2002, 364, 8-17.

[15] Andersson, T.; Nilsson, K.; Sundahl, M.; Westman, G.; Wennerström, O. J. *Chem. Soc. Chem. Commun.* 1992, 604-606.

[16] Sundahl, M.; Andersson, T.; Nilsson, K.; Wennerström, O.; Westman, G. *Synth. Met.* 1993, 56, 3252-3257.

[17] Hungerbühler, H.; Guldi, D. M.; Asmus, K.-D. *J. Am. Chem. Soc.* 1993, 115, 3386-3387.

[18] Yamakoshi, Y. N.; Yagami, T.; Fukuhara, K.; Sueyoshi, S.; Miyata, N. J. *Chem Soc. Chem. Commun.* 1994, 517-518.

[19] Scott, G. D.; Charlesworth, A. M.; Mak, M. K. *J. Chem. Phys.* 1964, 40, 611-612.

[20] Scott, G. D.; Kilgour, D. M. Br. *J. Appl. Phys.* 1969, 2, 863-866.

[21] Baram, A.; Luban, M. J. *Phys. C Solid Satate Phys.* 1979, 12, L659-L664.

[22] Alievsky, D. M.; Kamenin, I. G.; Kadushnikov, R. M.; Alievsky, V. M. *Modelling* 2001, 1, 1-3.

[23] Torrens, F. In Functional Nanomaterials, Geckeler, K. E.; Rosenberg, E.; Eds.; American Scientific: Stevenson Ranch (CA), in press.

[24] Torrens, F. In: Frontiers in Drug Design and Discovery I, Caldwell, G. W.; Atta-ur-Rahman; Springer, B. A.; Eds.; Bentham: Hilversum (Holland), 2005; pp 231-266.

[25] Torrens, F. Internet Electron. *J. Mol. Des.* 2004, 3, 514-527.

[26] Torrens, F. Internet Electron. *J. Mol. Des.* 2005, 4, 59-81.

[27] Torrens, F. *Int. J. Quantum Chem.*, in press.

[28] Torrens, F. *Mol. Simul.* 2005, 31, 107-114.
[29] Torrens, F. *J. Mol. Struct.* (Theochem), in press.
[30] Torrens, F. *Nanotechnology* 2005, 16, S181-S189.
[31] Torrens, F. Probl. Nonlin. *Anal. Eng. Syst.* 2005, 11(2), 1-16.
[32] Torrens, F. Molecules, submitted for publication.
[33] Torrens, F.; Castellano, G. *Comput. Lett.*, submitted for publication.
[34] Torrens, F.; Castellano, G. *J. Phys. Conf. Ser.* in press.
[35] Torrens, F.; Castellano, G. Phys. Status Solidi, submitted for publication.
[36] Bezmel'nitsyn, V. N.; Eletskii, A. V.; Stepanov, E. V. *In Progress in Fullerene Research*, Kuzmany, H.; Ed.; World Scientific: Singapore, 1994; pp 45-45.
[37] Bezmel'nitsyn, V. N.; Eletskii, A. V.; Stepanov, E. V. *J. Phys. Chem.* 1994, 98, 6665-6667.
[38] Bezmel'nitsyn, V. N.; Eletskii, A. V.; Stepanov, E. V. *Zh. Fiz. Khim.* 1995, 69, 735-735.
[39] Bezmel'nitsyn, V. N.; Eletskii, A. V.; Okun', M. V. *Physics–Uspekhi* 1998, 41, 1091-1114.
[40] Hotelling, H. *J. Educ. Psychol.* 1933, 24, 417-441.
[41] Hubert, M.; Engelen, S. *Bioinformatics* 2004, 20, 1728-1736.
[42] Tryon, R. C. *J. Chronic Dis.* 1939, 20, 511-524.
[43] Jarvis, R. A.; Patrick, E. A. *IEEE Trans. Comput.* 1973, C22, 1025-1034.
[44] McGregor, M. J.; Pallai, P. V. *J. Chem. Inf. Comput. Sci.* 1997, 37, 443-448.
[45] Doman, T. N.; Cibulskis, J. M.; Cibulskis, M. J.; McCray, P. D.; Spangler, D. P. *J. Chem. Inf. Comput. Sci.* 1996, 36, 1195-1204.
[46] Turner, D. B; Tyrrell, S. M.; Willett, P. *J. Chem. Inf. Comput. Sci.* 1997, 37, 18-22.
[47] Reynolds, C. H.; Druker, R.; Pfahler, L. B. *J. Chem. Inf. Comput. Sci.* 1998, 38, 305-312.
[48] *Integrated Mathematical Statistical Library (IMSL),* IMSL, Houston, 1989.
[49] Page, R. D. M. Program TreeView, Universiy of Glasgow, 2000.
[50] Huson, D. H. *Bioinformatics* 1998, 14, 68-73.
[51] Hocking, R. R. *Biometrics* 1976, 32, 1-49.
[52] He, L.; Jurs, P. C. *J. Mol. Graphics Mod.* 2005, 23, 503-523.
[53] Smirnov, B. M. *Usp. Fiz. Nauk* 1997, 167, 1169-1169.

In: Focus on Nanotube Research
Editor: Delores A. Martin, pp. 67-96
ISBN: 1-59454-851-X

Chapter 4

MICRO-TOPIARY – LASER PRUNING OF CARBON NANOTUBES ARRAYS FABRICATION OF STATIC AND MOVABLE CNTS STRUCTURES VIA LASER TRIMMING

Chorng-Haur Sow *

Department of Physics, Faculty of Science, 2 Science Drive 3,
National University of Singapore, Singapore 117542;
National University of Singapore Nanoscience and Nanotechnology Initiative,
2 Science Drive 3, Singapore 117542

Kim-Yong Lim, Fook-Chiong Cheong, and Neelam Saurakhiya

Department of Physics, Faculty of Science, 2 Science Drive 3,
National University of Singapore, Singapore 117542

Abstract

We report our efforts in developing laser beams as simple means to fabricate patterned arrays of aligned CNTs or three-dimensional structures made of CNTs without a pre-patterned substrate. Two related techniques were developed. The first technique makes use of a tightly focused laser beam from a moderate power He-Ne laser as precision trimming tool to create unique 3D CNTs structures. The details of the technique will be presented. The technique can be used to modify the surface morphology of the CNTs forest. The technique also allows for a wider range of three-dimensional CNTs microstructures to be constructed. In addition, some of the structures created were found to exhibit laser-beam induced mechanical motion. Such behavior can be exploited in creating possible micro-optomechanical devices made of CNTs. We demonstrated the feasibility of a simple optomechanical switch with this method. The second technique makes use of a parallel or converging laser beam from a high power UV pulsed laser for the destruction of aligned CNTs with a copper grid as lithography mask to define patterned aligned CNTs arrays. This technique is simple to implement, inherently

---

* E-mail address: physowch@nus.edu.sg (Corresponding author)

parallel and has high throughput. It is well suited for creating a large extended area of patterned CNTs array.

A Introduction

Carbon nanotubes (CNTs) have been of great interest to researchers in recent years due to their remarkable structural, electronic and mechanical properties [1]. As they have tunable electronic properties, they are being investigated as possible new materials for the next generation of microelectronic and nanoelectronics devices [2]. They have also been identified as potentially useful materials for a broad range of functions [3,4] including actuator [5], data storage [6], fuel cells [7], flat panel displays [8, 9,10,11] biosensors [12] and improved X-ray sources [13]. Arrays of CNTs have also been used as templates for a variety of materials to achieve a variety of functions such as creating super-hydrophobic surfaces [14]. On the other hand, such an array of CNTs can be used as template on which other nanomaterials can grow on. [15] Many of these applications require a fabrication method capable of producing extended areas of patterned and aligned CNTs with uniform structures and periodic arrangements to meet device requirements. As an example, there has been a recent attempt to create complementary logic and ring oscillator circuits from arrays of CNTs-based transistors and inverters [16].

There has been continuing efforts in micropatterning of CNTs. Several successful attempts have been made to create the patterned periodic array of CNTs onto various substrates [17,18]. Typical patterning methods often involve conventional lithographic technique resulting in a pre-patterned substrate on which the arrays of CNTs can be grown. There are a few conventional methods of building patterned aligned CNTs arrays. One is by means of growth on a patterned catalyst layer whereby the surface of a planar substrate is first covered with a mask and then deposited with catalytic nanoparticles from which the nanotubes can grow [3]. Other techniques make use of electron beam lithography to define the patterns of catalytic nanoparticles for the growth of CNTs [19]. A more recent development made use of substrate etching mechanisms to define the arrays' directional growth [20]. Yet, another group made use of self-assembled polystyrene spheres to act as a mask during deposition of the catalytic layer [21].

Such fabrication methods often involve multiple-step pretreatment of the substrate before successful growth of arrays can be implemented. Also, such methods are only feasible with patterning in 2 dimensions. To realize three-dimensional nanotube-based devices, patterning in the third dimension is required. Even in the simple case of having CNT arrays of varying heights on the same substrate, the task becomes formidable with conventional lithographic techniques or *in situ* growth techniques. Recent advances in this direction made use of nanotube growth to lift up a metal film to achieve the link-up between the planar substrate and thin films in the vertical direction. Self-directed growth of CNTs between nanoscale silicon pillars resulting in a 3-dimesional CNT network has also been achieved [22]. However, the structures fabricated by these techniques do not always establish the desired length, direction and pattern control of the CNTs array.

In this report, we present our efforts in the development of two related techniques that facilitate the creation of unique CNTs structures on a sample consisting of uniformly-length CNTs grown on Si substrates. These techniques share the advantage of being able to generate arrays of patterned CNTs without a pre-patterned substrate. In addition, the controlled fabrication of CNTs arrays of different lengths on the same substrate is demonstrated. The first technique built on the laser-pruning technique that we have reported earlier in reference [23]. In brief, the first technique involves the use of a tightly focused laser beam from a moderate power He-Ne laser as precision burning tools to create unique 3D CNTs structures. The technique can be used to modify the surface morphology of the CNTs forest. The technique also allows for a wider range of three-dimensional CNTs microstructures to be constructed. In addition, some of the structures created were found to exhibit laser-beam induced mechanical motion. Such behavior can be exploited to create possible micro-optomechanical devices made of CNTs. Here, we demonstrate a simple switch that is controlled by focused laser beam. The second technique also built on another technique that we have reported earlier [24]. It makes use of a parallel or converging laser beam from a high power UV pulsed laser for the destruction of aligned CNTs with a copper grid as lithography mask to define patterned aligned CNTs arrays. This technique is simple to implement, inherently parallel and has high throughput. It is well suited for creating a large extended area of patterned CNTs array and is useful in field emission device application by way of patterning as well as variation of the density of the CNTs.

This report is organized as follow. After the introduction in section A, we proceed with a discussion of the Plasma Enhanced Chemical Vapor Deposition (PECVD) techniques for the growth of CNTs in Section B. All the samples used in this report were fabricated using the described technique and recipes. This is followed Section C where we provide a detail description of the experimental setup of our focused laser system that we used to trim a forest of CNTs. We will first demonstrate the feasibility of this simple and efficient technique in trimming a forest of CNTs. Other aspects of the technique such as the dependence of the resultant CNTs structures on the profile of the focused laser beam and the effects of laser pruning with an oblique beam will be presented. Section D demonstrates some of the possible applications of the laser pruning method. Section E introduces a high throughput method of laser pruning using a highly energetic pulse laser and this is followed by a discussion on its applications.

B PECVD Growth of Aligned CNTs Forest

Carbon nanotubes of uniform height can be synthesized via a number of different techniques [1]. In this work, we employed Plasma-Enhanced Chemical Vapor Deposition System (PECVD) with the assistance of iron catalyst nanoparticles. Firstly thin iron films were deposited on silicon or quartz substrate by means of rf magnetron sputtering. Si substrates that were sputtered with iron film were then placed in the PECVD unit for the growth of CNTs. Since the iron films were exposed to atmosphere, it could become oxidized in atmosphere. Therefore it was first subjected to hydrogen plasma for 600 seconds at 750 °C in PECVD unit to reduce the film to metallic iron nanoparticles. Subsequently, a mixture of C_2H_2 and H_2 gases with a flow rate of 15 sccm and 60 sccm respectively were then introduced at a temperature of 750°C, a pressure of 1200mtorr and a rf plasma of 100W.

Fig. 1 SEM image of the side view of aligned and uniformly length CNTs forest grew perpendicularly on the Si substrate. The length of the CNTs forest is about 95 μm. The scale bar is 20 μm.

Alternatively, we can use a different recipe consisting of a mixture of C_2H_2 and NH_3 for the synthesis of CNTs. In this case, the sample was first treated by NH_3 for 240 seconds with the flow rate of 30 sccm while the substrate was maintained at a temperature of 750 °C. A mixture of C_2H_2 and NH_3 is then introduced at a flow rate of 30 sccm and 15 sccm respectively. The pressure was kept at 1000mtorr throughout the process. Depending on the lengths of CNTs required, the time of CNTs growth was varied accordingly. Typically, for the former recipe, 10μm of CNTs will require 10 minutes of growth time. The result was a large area of aligned CNTs arrays on quartz or silicon. The arrays of CNTs grew perpendicular to the surface, were of uniform length and were typically about 30nm in diameter. Fig. 1 shows a Scanning Electron Microscope (SEM) image of the side view of aligned CNTs forest with a uniform length of ~ 95 μm grown on Silicon substrate. The image was taken using a JEOL JSM-6400F field emission scanning electron microscope system operating at 5kV. In this work, we have made use of samples of CNTs with different lengths ranging from 1μm to 100μm.

C Experimental Technique I

C.1 Experimental Setup: Focused Laser Beam

Fig. 2 shows a schematic of the experimental set-up of optical microscopes- focused laser beam system utilized in this work. The system consists of two optical microscopes (optical microscope A and optical microscope B) and a He-Ne laser source with moderate power. The role of microscope A was to focus the laser beam onto the sample and at the same time capture the image of the sample. The role of microscope B was to capture the laser pruning process from a side view. The laser used was a Coherent He-Ne laser (λ=632.8nm) emitting parallel beam (divergence of 0.66 mrad) with a beam width of 1.25 mm and a measured output beam power of 38 mW. The He-Ne laser beam was directed into the microscope A via two reflecting mirrors (M1 and M2). Inside microscope A, the laser beam was directed towards an objective lens (L) via a beam splitter (S1). A second (and optional) beam-splitter (S2) was used for directing illumination light from an illumination lamp that is part of the microscope system. Alternatively, illumination of the sample can be achieved via an external Olympus LPGS light source (not shown in Fig. 2) without engaging S2. The power of the laser beam, after passing through the M1-M2-S1 setup, was measured to be 29 mW. The laser beam was then focused by the objective lens (L). During the cutting of the CNTs, the same objective lens (L) was used to collect reflected light from the sample for viewing purposes. The reflected optical image was captured by a CCD camera (CL350 Panasonic CCD camera) that was coupled to a monitor. In this way, we could inspect the structures created and move the lens to focus the laser beam to different levels.

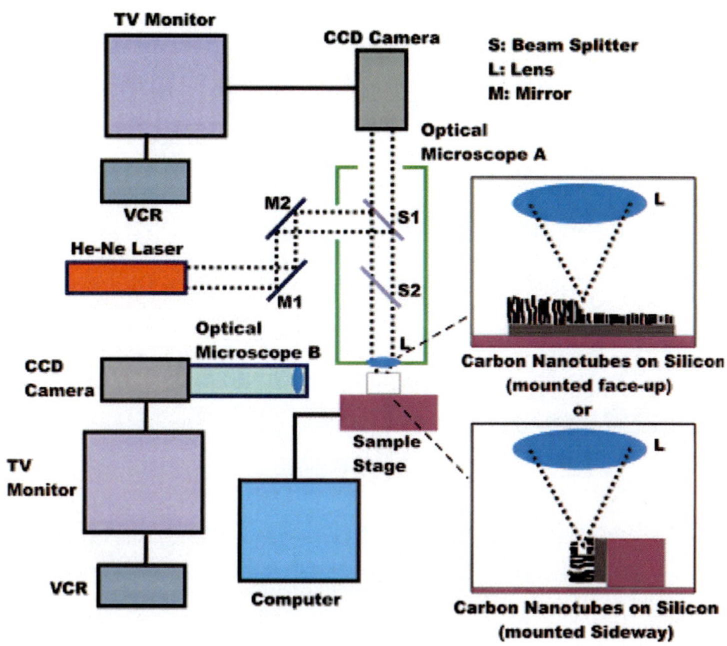

Fig. 2 Schematics of the optical microscopes-focused laser beam system used in this work.

In this work, we have made use of two different objective lenses to focus the laser beam onto the samples. The first objective lens used was an Olympus MDPlan 150× lens with numerical aperture of 0.95. The second objective lens used was a Nikon 50X lens with a numerical aperture of 0.55 and a long working distance of 8.7 mm. The sample was placed on a computer-controlled, motorized X-Y stage. The sample stage used was a DynaOptic Motion CTC-512-2 XY stage with 25mm travel and a minimum step-size of 50nm in the x and y directions. It was controlled by a CTC-290-10X motor controller and interfaced to the computer via the Motion Control API software. As shown in Fig. 2, microscope B with a long working distance was added to the system so that during experiments, both the top view and the side view of the sample can be imaged simultaneously. The long working distance microscope was coupled to a JVC KY-F50 camera.

When the focused laser beam was incident upon the CNTs sample, it readily trimmed away the top layers of the CNTs. SEM images revealed that the part of the CNTs was removed cleanly without leaving behind any residue. The length of the CNTs that was removed depends on a number of factors that will be discussed in later sections. The detailed mechanism of the laser pruning process is still under investigation. It is possible that the CNTs absorbed the energy from the laser beam and this result in the disintegration of the CNTs into active constituents that escape into the region above. We found that the laser pruning worked even if the CNTs samples were kept in a vacuum of 10^{-2} torr. With the computer control stage, we can move the sample with respect to the focused laser beam and cut out any desired patterns onto the forest of aligned CNTs. In addition, as shown Fig. 2, the CNTs samples could be mounted either in a face-up manner or a sideway manner. In this way, we could make unique three-dimensional (3D) structures made of CNTs. This process bear a close resemblance to pruning of plants into unique structures, hence we call this simple laser pruning method the Micro-Topiary technique.

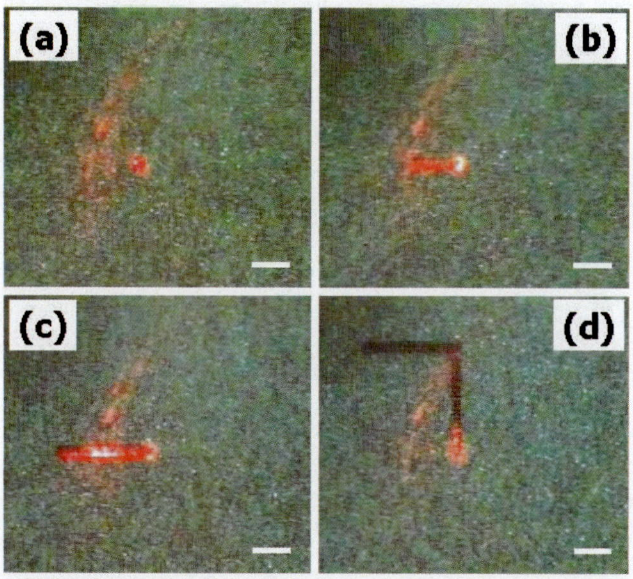

Fig. 3 (a-d) Sequential optical microscope images of the top view of the CNTs forest trimmed with a focused laser beam. These sequential images show the clean removal of the CNTs with the focused beam. Scale Bar = 20μm. Videoclips of the laser pruning process can be found at the following website [25].

Fig. 3 shows sequential images of the top view of a CNT sample during laser pruning process. The images were obtained using optical microscope A (see Fig. 2) and the laser beam was focused by a 50X lens. Fig. 3 shows the clean removal of the CNTs as the focused laser beam scans through the sample. Videoclips of the laser pruning process can be found at the following website [25]. Fig. 4(a-c) shows sequential images of the side view of a CNT sample during laser pruning process. The images were obtained using optical microscope B (see Fig. 2). In this case, the laser beam was focused by a 50X lens. With this lens, the laser beam is less tightly focused and this facilitates the trimming of sharp trenches in the aligned CNTs forest. Such precision cutting capability provides an important means to create 3D structures made of CNTs.

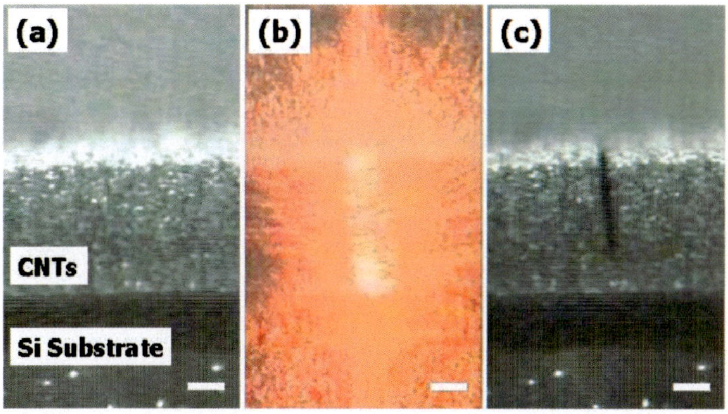

Fig. 4 Optical micrographs of the side view of aligned CNTs profile during the laser trimming process as capture by optical microscope B (see Fig. 2). The images show (a) side view of the aligned CNTs forest; (b) incident of laser beam and (c) a sharp trench indicating the removal of CNTs. Scale bar = 10 μm.

By using samples with relatively short CNTs, the focused laser beam removed the CNTs completely and revealed the underlying substrate. Direct writing of the CNTs arrays with the focused laser beam resulted in two-dimensional (2D) structures made of CNTs. One could fabricate two types of patterns. Negative patterns are those where the desired patterns (either word or figures) are voids previously occupied by CNTs. Positive patterns are those where the desired patterns are made up of CNTs themselves. They are created by removing CNTs such that the patterns comprising of CNTs become discernable. Examples of positive patterns are shown Fig. 5 where the letters (a) 'NUS' and (b) 'NOVA' are comprised of CNTs when the CNTs surrounding the letters were removed.

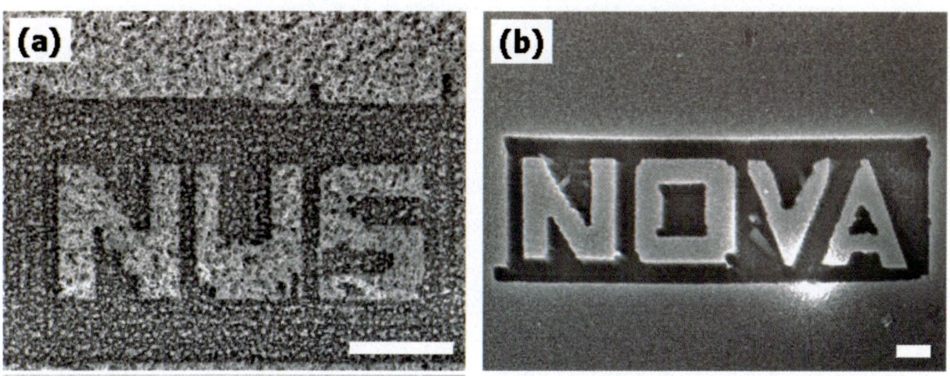

Fig. 5 SEM images of positive CNTs patterns. The letters (a) 'NUS' and (b) 'NOVA' were made of CNTs. They were created by removing the CNTs surrounding the letters with the laser beam. Scale bars (a) 10 μm (b) 20 μm.

In addition to 2D structures, we can also make use of the laser pruning technique to create 3D structures made of CNTs. One of the advantages of the simple laser pruning technique is that it can be utilized in different methods in order to create 3D structures made of CNTs. To construct the 3D structures, typically CNTs arrays with longer tubular lengths (>

10 μm) were used. In one method, we made use of the observation that the level of the trimmed CNTs depends on the focus level and power of the laser beam. Using a moderate laser power, we first focused the laser onto the surface and removed a thin layer of CNTs from the top, after which we moved the objective lens closer to the sample by focusing onto the freshly cut layer. Another thin layer of a different pattern was scanned and removed at this level. This process was repeated successively until the desired structures were created. Examples of 3D CNTs structures fabricated using this method are shown in Fig. 6(a, b). Alternatively, one can simply vary the length of trimmed CNTs by changing the laser power. On the hand, one can also fabricate 3D CNTs structures by a two-steps process where the CNTs samples were first mounted and trimmed in a "face-up" manner followed by mounting and trimming in a "side-way" manner (see Fig. 2). Fig. 6(c) shows an example of unique "π" shape 3D CNTs structures created in this way.

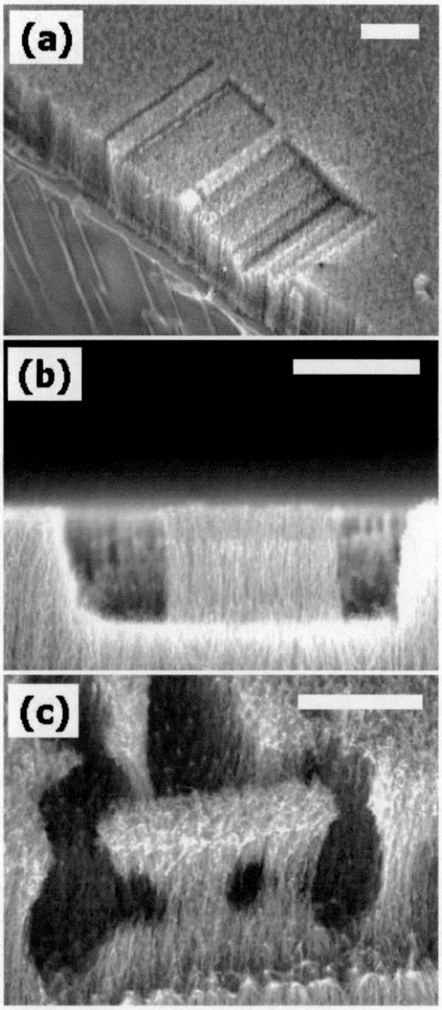

Fig. 6 SEM images of (a) 3D Staircase-like structures cut out from the CNT forest. (b) Single CNTs platform (c) 3D "π" shape CNTs structures. Scale bars: 10 μm.

To determine the limits imposed on this technique, an attempt was made to fabricate a small flap of CNTs by trimming it down. Fig. 7 shows a SEM image of the resultant flap of CNTs. The width (as the base) of the CNT flap is about 650nm. This limitation is due to the optical microscopy as it is required that the flap remain visible while it is being trimmed down.

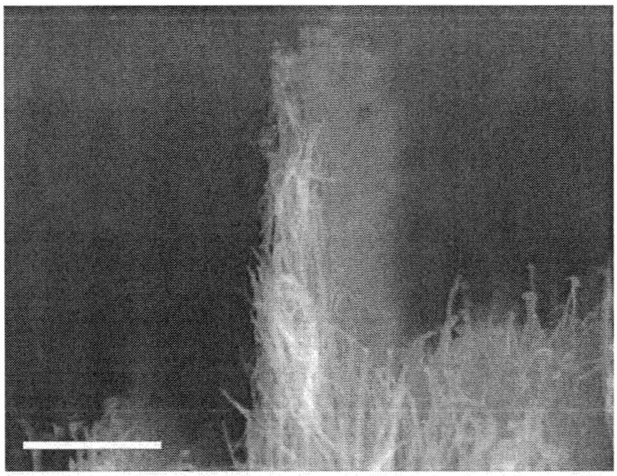

Fig. 7 SEM image of a small flap of CNTs. Scale bar is 1 μm.

The ease of creating structures of different scales is exemplified by how the depth and width of the laser cut varies with the magnification of the lens used. An objective lens with a magnification of 50X and long working distance will produce a deeper cut than an objective lens with a magnification of 150X by virtue of their different laser profiles. To further study the profile of the volume of CNTs removed by a converging laser beam, a variety of experiments were carried out. These experiments were carried out to explore the extent of control one has over the types of possible CNTs structures and to provide an idea of how the laser pruning of CNTs vary with parameters such as the scan speed of the laser beam, CNTs depth, alignment of laser spot, laser power and the distance of the lens from the CNTs surface.

C.2 Width of Laser Pruned CNTs Channels versus the Scan Speed of the Laser Beam

Using the objective lens with the magnification of 150X and a laser power of ~ 20mW, the laser beam was focused onto the top surface of a CNTs sample. Using the computer control stage, the CNTs sample was moved at constant speed with respect to the focused laser beam. The speed was varied in this experiment. This process resulted in the laser beam scanning across the surface of the CNTs sample and creating parallel channels. Fig. 8 shows a SEM image of the parallel channels created. It is evident that the width of the channel depends on the scanning speed of laser beam.

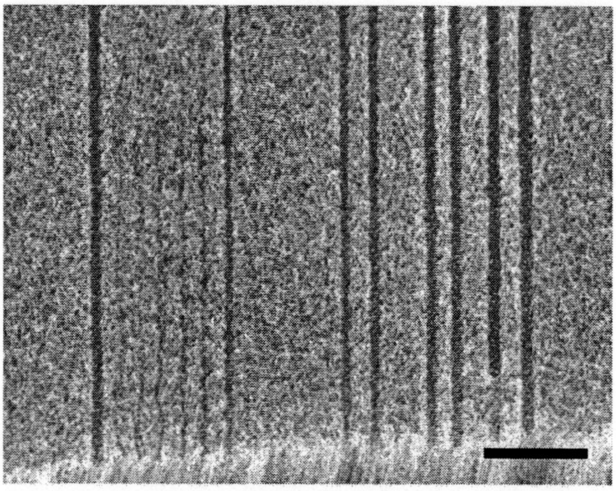

Fig. 8 SEM images of top view of CNTs sample with parallel channels fabricated with different scanning speed of the laser beam. Scale bar = 10 μm.

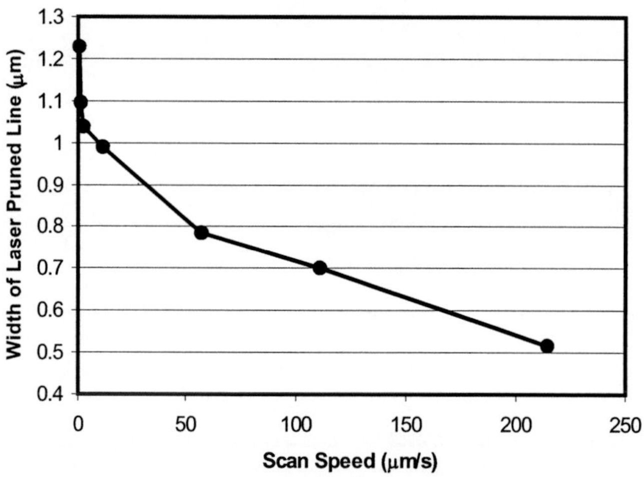

Fig. 9 Graph of the width of the laser pruned channel versus the scanning speed of the laser beam. The solid line is guide to eye.

Fig. 9 shows a plot of the width of the channel versus the scanning speed. As the scan speed was increased, the channel removed decreased in width. The widths of the channels were measured along several places along each scan line and averaged. A range of 0.5 to 1.5 μm was obtained for the width of the cut line. The threshold scan-speed for clear cutting of the array of CNTs was found to be at approximately 210μm/s. From the plot, we deduce that the profile of the volume trimmed by the laser beam depends sensitively on the exposure time of the CNTs to laser beam. The laser beam we used has a gaussian intensity profile. As the scan speed increases, only the more intense portion of the laser profile has the power to trim the CNTs at the given exposure time. Hence, the laser beam will trim less volume of the CNTs forest when the scan speed is high.

C.3 Laser Trimmed Profile versus Height of the Objective Lens

The results presented in the previous section indicated that the profile of the volume trimmed by the laser beam depends sensitively on the profile of the laser beam. Hence, by moving the objective lens closer to or further away from the CNTs sample, we would be subjecting the CNTs forest to different effective light cone of the converging laser cone. Fig. 10(a-c) show schematics of the CNTs sample and the focused laser beam profile at different position of the objective lens. In this case, Fig. 10(b) corresponds to the situation where objective lens was brought to focus onto the top surface of the CNTs forest. Indeed a clear image of the top view of the CNTs forest was observed. Thus this position of the objective lens was regarded as the reference position (the "0" relative position of the lens to the CNTs forest). If the lens is moved closer to the CNTs forest as shown in Fig. 10(a), we denote this situation as corresponding to a positive value of the relative position of the lens. Likewise, if the lens is moved away from the CNTs forest as shown in Fig. 10(c), we denote this situation as corresponding to a negative relative position of the lens.

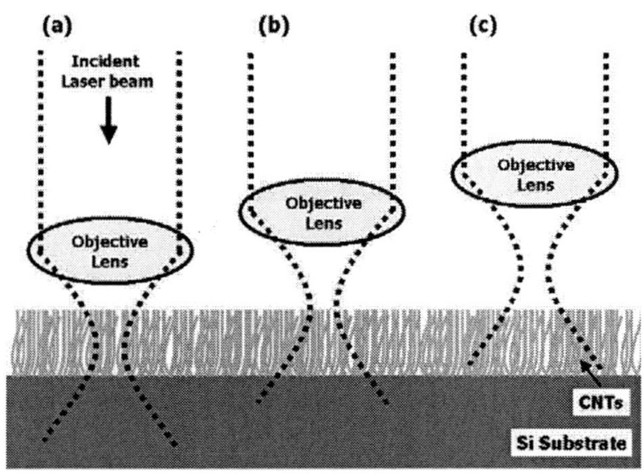

Fig. 10 Schematics of the CNTs sample and the focused laser beam profile corresponding to different positions of the objective lens.

As illustrated in Fig. 10, the profile of the trimmed CNTs forest will depend sensitively on the relative position of the lens. When the lens is away from the CNTs forest (Fig. 10(c)), the CNTs was under the influence of diverging light cone and hence would result is a wider cut. However, since the energy concentration of diverging light cone is smaller, that implies the laser cut would be shallow. When the lens focus onto surface of the CNTs forest (Fig. 10(b)), a narrow cut naturally occurs at the surface of the CNTs forest and the diverging light cone is sufficiently intense to remove more CNTs beneath the surface level. When the lens is close to the CNTs forest, the laser trimmed profiles become very interesting. They reflect the effects of the interplay between the converging-diverging light cones with the effective laser power in the light cone. Fig. 11 show SEM images of the side profile of a CNTs sample showing a range of different laser trimmed profiles. They are created by systematically vary the relative position of the objective lens (with a magnification of 150X). In Fig. 11(a), the two right most laser-trimmed profiles correspond to case where the lens is focused onto the top surface of the CNTs sample (i.e. "0" relative position.) The laser-trimmed profiles to the left of these correspond to the resultant profile when the lens was moved systematically closer to the CNTs forest at a step size of 5 μm (i.e. positive relative position of the lens). Similarly, in Fig. 11(b), the two left most laser-trimmed profiles correspond to "0"-position and the profiles to the right corresponds to the resultant profile when the lens was moved systematically away from the CNTs forest at a step size of 5 μm (i.e. negative relative position of the lens).

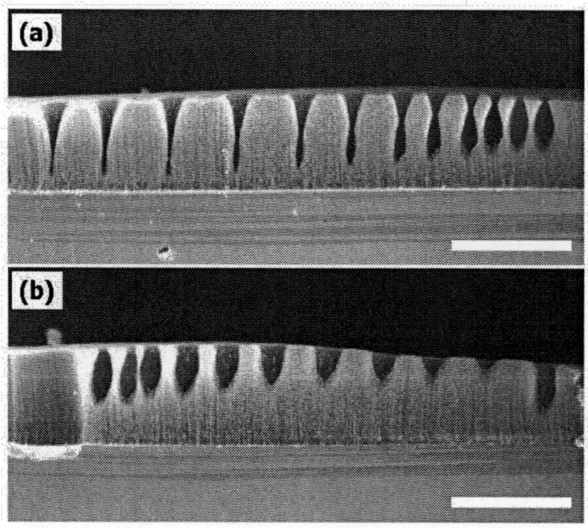

Fig. 11 SEM images of the side profile of aligned CNTs forest showing the laser trimmed CNTs profile fabricated as a result of different position of the lens (150X) w.r,t. the CNTs sample. Scale bars: 100 μm.

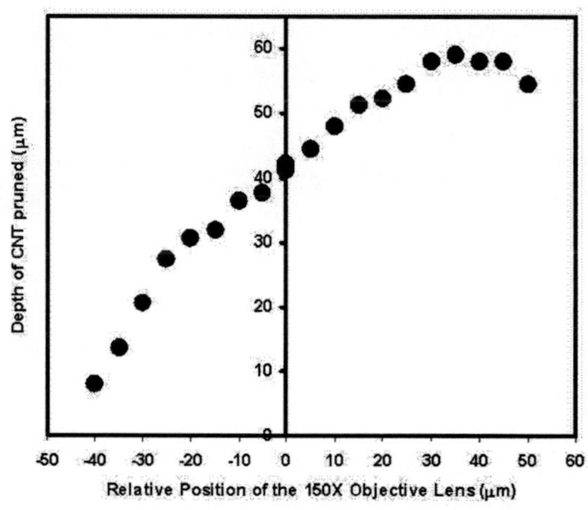

Fig. 12 Trimmed CNTs depth as measured from the top level of the CNT forest profile versus the position of the lens (with a magnification of 150X).

Fig. 12 shows a plot of the measured depth of laser-trimmed profiles from Fig. 11 versus the position of the lens. From Fig. 12, for small to moderate distances from the "0" position, the relationship between the depth of the laser cut and relative position of the lens is approximately linear. For larger positive values of relative position of the lens, the extent of the depth cut out by the laser levels off. As the laser focal spot lies deeper inside the CNTs, the laser power is absorbed or scattered by the surrounding CNTs on its way to the focal point. Hence, the remaining laser power will not penetrate as deeply as where the laser is allowed to arrive at the focal point in a free path.

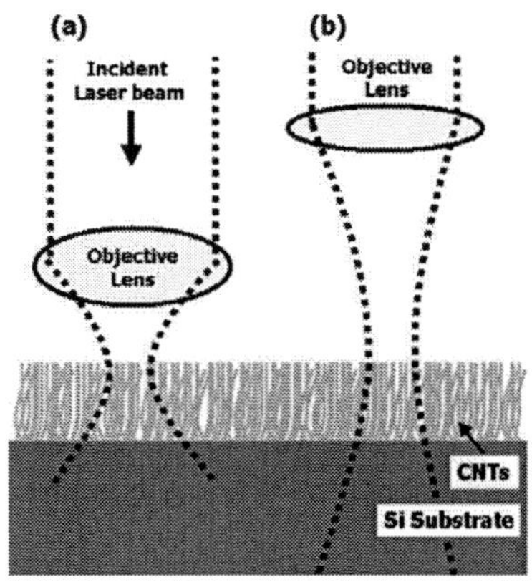

Fig. 13 Schematics of the CNTs sample and the focused laser beam profile corresponding to objective lens with different magnifications and numerical apertures.

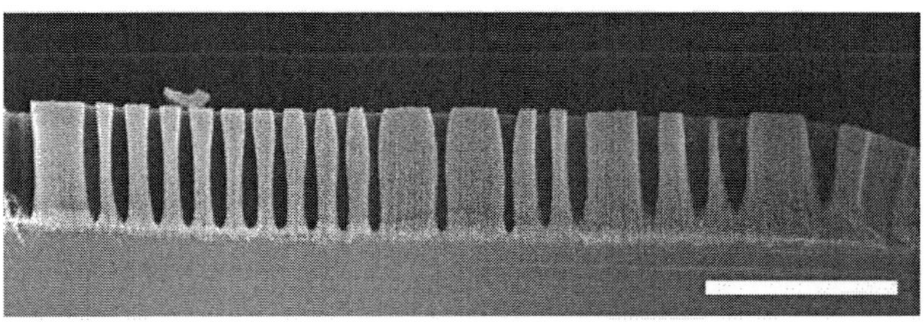

Fig. 14 SEM image of side profile of aligned CNTs forest showing laser trimmed CNTs profile fabricated as a result of different position of the lens (50X) w.r.t. the CNTs sample. Scale bar: 100 μm.

The laser-trimmed profiles of objective lens with different magnification were also investigated. In this case, we replaced the 150X lens with an objective lens with a long working distance and a magnification of 50X. Similar to before, the lens was positioned as different position with respect to the CNTs forest and the resultant laser-trimmed profiles were imaged. As shown in the schematic in Fig. 13, such a lens would yield a less converging focused light cone and hence the profiles of the laser-trimmed region are expected to yield sharper edges. Fig. 14 shows SEM image of the side profile of the resultant laser-trimmed profile obtained in this experiment.

C.4 Depth of Laser Trimmed Profile versus Laser Power

We have also studied the relationship between the depth of the laser-trimmed profile and the power of the laser beam. In this experiment, we made use of the 150X-lens and positioned the lens such that laser beam focused onto the top surface of the CNTs forest (i.e. the "0"-

position). After which the CNTs was pruned by using different laser power in the range from 9 mW to 29 mW (laser power is measured before entering the microscope system). The power of the laser was varied by placing a polarizer in the path of the polarized laser beam before it entered the microscope setup Fig. 15(a) shows SEM image of the side profile of aligned CNTs forest showing laser trimmed CNTs profile created in this experiment. Fig. 15(b) shows a plot of the measured depth as a function of the power of the laser beam. The depth of the laser-trimmed regions varies linearly with the power of the laser beam. A threshold of ~ 8 mW was required to cause any significant laser pruning. Fig. 15 thus provides a calibration that allows one to fabricate CNTs with the desired length. It is interesting to note that despite the difference in the depth of laser-trimmed regions, the shape of the laser-trimmed profile remained similar as this was mainly due to the converging light cone that remained the same in this series of experiments.

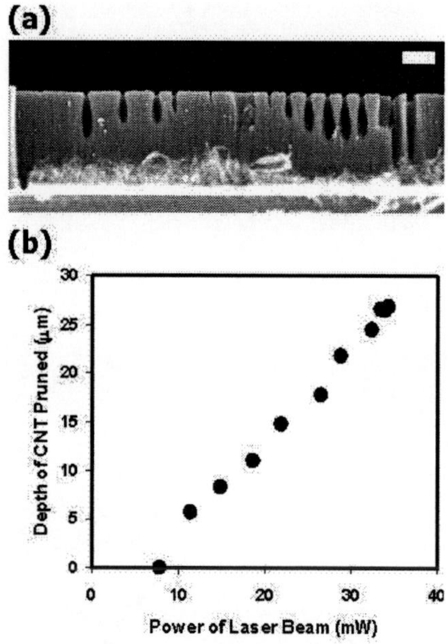

Fig. 15 (a) SEM image of the side profile of a CNTs forest trimed with focused laser beam with different power. Scale bar: 20 μm. (b) A plot of the depth of the CNTs trimmed versus the measured power of the laser beam.

C.5 Laser Trimming Using an Oblique Beam

Apart from the control of profiles of the laser cut and the depth of cut with position of the lens and the power of the laser beam, the angle of cut into the CNTs arrays could also be varied. This was achieved by intentionally adjusting the mirror (M2, see Fig. 2) that serve to reflect the laser beam into the microscope. Adjustment of the mirror resulted in the laser beam incident on the CNTs at an oblique angle, as illustrated in Fig. 16. As a result, one could trim the CNTs obliquely. Fig. 17(a) shows a mosaic of SEM images of the side view of a CNTs forest showing the laser-trimmed profiles using laser beam at different oblique angle.

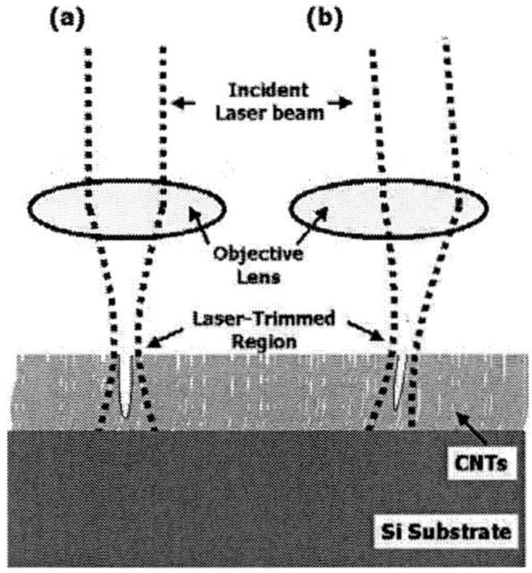

Fig. 16 Schematics illustrating the different of the laser trimmed profile due to (a) an aligned beam and (b) an obliquely incident laser beam.

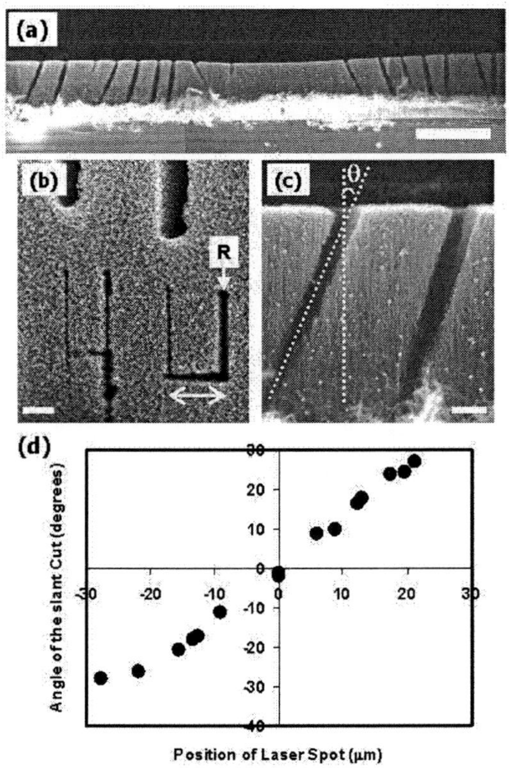

Fig. 17 (a) A mosaic of SEM images of the side view of a CNTs forest showing the laser-trimmed profiles using laser beam at different oblique angle. Scale bar: 100μm. (b) SEM image of the top view of two channels cut by aligned beam marked R) and oblique beam. The double-head arrow denotes the "position of the laser spot". Scale bar: 10 μm. (c) SEM image of the side view of the obliquely trimmed CNTs profile corresponding to Fig. (c). The angle, θ, marks the slant-cutting angle. Scale bar: 10μm. (d) A plot of the slant-cutting angle θas a function of the "position of the laser spot".

During the process of laser cutting, the CNTs forest was viewed from the top using optical microscope A. Hence one requires a way to estimate the cut angle based on some observable parameter. This was done by the following procedure. First we identified the position where the laser is at normal incidence to the surface of the sample. This corresponds to a straight cut where the laser-trimmed profile is normal to the surface of the sample. Using this beam profile, a short CNT channel was cut as a reference. As the mirror was adjusted, the laser ray became oblique, the position of the focus spot as observed on the surface of the CNTs forest shifts away from the reference. A second CNTs channel was trimmed all the way to the edge of the sample. The distance of the second channel from the first one was denoted as the "position of the laser spot". After laser trimming, both the top view and side view of the sample were imaged via SEM. Figure 17(b) and (c) shows the top view and side view of some channels created in this way. From the SEM images, we measured the "position of the laser spot" and the corresponding tilting angle, denoted as "slant-cutting angle", θ, of the resultant obliquely trimmed CNTs profile. Fig. 17(d) shows a plot of the slant-cutting angle, θ as a function of the "position of the laser spot". Such a plot is a useful calibration line for our system.

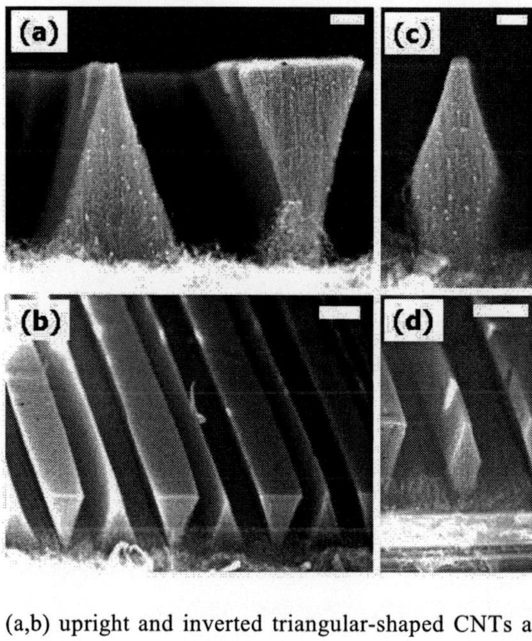

Fig. 18 SEM images of (a,b) upright and inverted triangular-shaped CNTs arrays and (c,d) diamond-shape CNTs array fabricated using the obliquely incident laser beam. Scale bars: (a,c) 10 μm, (b,d) 50 μm.

We can exploit this capability to make unique 3-D structures without tilting the sample. For example, Fig. 18(a-b) show both upright and inverted triangular-shaped CNTs arrays fabricated using the obliquely incident laser beam. It should be noted that the cross sections of CNTs arrays shown in Fig. 18(a-b) resemble that of an isosceles triangle. Using the calibration line (Fig. 17(d)), we were able to control the resultant angle of the slant cut into the CNTs forest and thus were able to create structures with symmetrical profiles. Besides the triangular-shaped CNTs array, we were also able to create a diamond-shaped CNTs arrays

(Fig. 18(c-d)) by cutting the same CNTs region twice each with different oblique laser beams. As mentioned before, one could fabricate unique 3-D CNTs structures via a two-steps process involving mounting the sample on the side. However, this method is not feasible for creating structures in the middle of the sample. The slant-cutting method using oblique laser beams provides a feasible solution to create unique 3D CNTs structures in the middle of the sample. Fig. 19 shows SEM images of a periodic array of pyramids cut out by the technique described above. The sample was first cut in one direction using oblique laser beams and then followed by cuts in the direction perpendicular to the former cut direction.

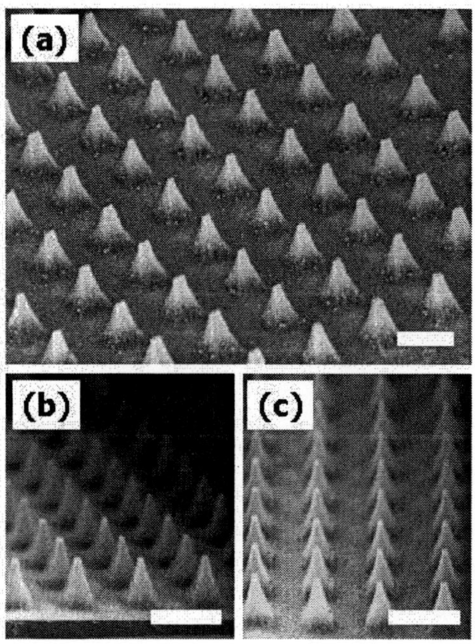

Fig. 19 SEM images of the periodic array of 3D CNTs pyramids fabricated by obliquely focused laser beam. Scale bars: 50 μm.

The creation of 3-D structures by controlling the different parameters, as explained before, provides a complementary technique to the many growth and bottom–up methods available. This combination of bottom-up and top-down approach allows 3-D architectures of CNTs to be created with relative ease. Hence, all these approaches provide feasible openings towards possible 3-dimensional electronic circuits that employ CNTs as interconnects or even as the electronic devices themselves.

D Applications OF Laser Pruning Technique

D.1 Morphology Modification of the Surface of the CNTs Forest

The effect of laser cutting also resulted in very different surface structures of the CNTs-covered sample. Fig. 20 shows SEM images of (a) surface of as-synthesized CNTs sample and (b) surface after a thin layer (~1μm) of the CNTs has been removed using a low power laser. Fig. 20(b) shows clear difference in surface morphology after laser treatment. The laser-trimmed region consisted of CNTs that were naturally shorter but more uniform and aligned. Sharp needle-like pointed CNTs were found together with numerous fused junctions consisting of two or more CNTs joined together. Raman spectroscopy studies of the as grown region and the laser-trimmed region indicated that the CNTs at the laser-trimmed region has lower degree of structural disorder [23] It is possible that the surface layer of the CNTs forest comprised of CNTs with more structural disorders. After the removal of the top layer of CNTs by the laser beam, the degree of structural disorder became reduced. Fig. 21 shows an SEM of the close-up view of the laser-trimmed sample surface.

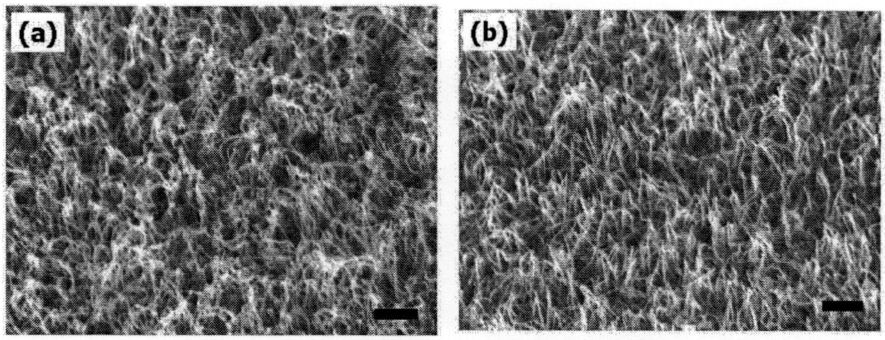

Fig. 20 SEM images of angled (25-degree) view of (a) surface morphology of the as-synthesized carbon nanotubes and (b) top surface of a CNTs forest where a thin layer of CNT was removed with laser beam. Scale bar: 1 μm.

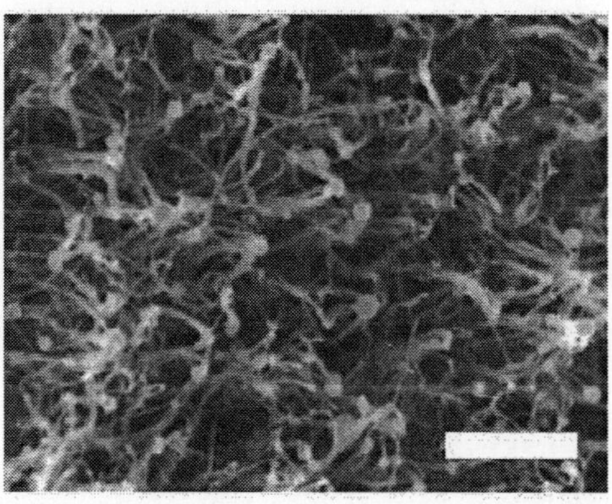

Fig. 21 SEM images of close-up view of fused junction and pointed CNTs needles. Scale bar: 1 μm.

To further illustrate that the CNTs were indeed fused together, we imaged the same group of CNTs under SEM before and after laser cutting. Fig. 22(a) shows the side-view of as grown CNT and Fig. 22(b) shows the result of cutting the CNTs at the sample surface using a 13mW laser beam. Here, the nanotubes indicated by the white arrows have changed their initial directions of alignments and merged at their tips to form a fused junction. This phenomenon has also been observed in the other fused junctions. Fig. 23 show images of fused junctions obtained using a Transmission electron Microscope (TEM) (Model JOEL JEM-3010). It clearly shows that the CNTs strands became fused together. We have observed junctions formed by as many as five CNTs. The detailed mechanism of the formation of such fused junction is still under investigation. It is possible that the laser trimmed CNTs comprise of active ends that become fused together under the influence of focused laser beam. Such capability could potentially be useful in joining CNTs together into some functional devices. It is also interesting to study the electrical transport properties of similarly fused CNTs.

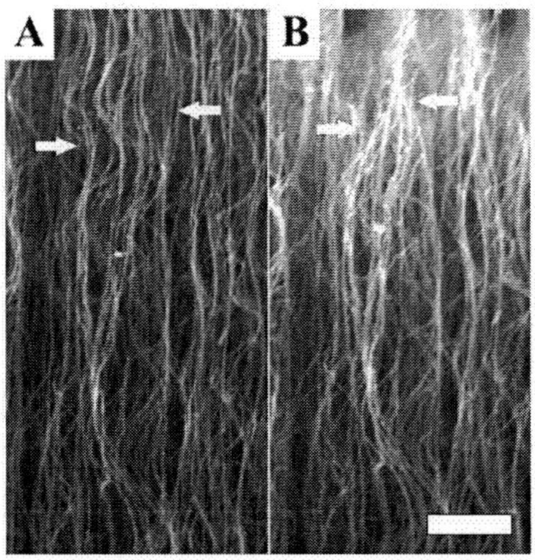

Fig. 22 SEM image of the side view of CNTs strands before and after the laser treatment. One can identify the fused portion. Scale bar = 1 μm

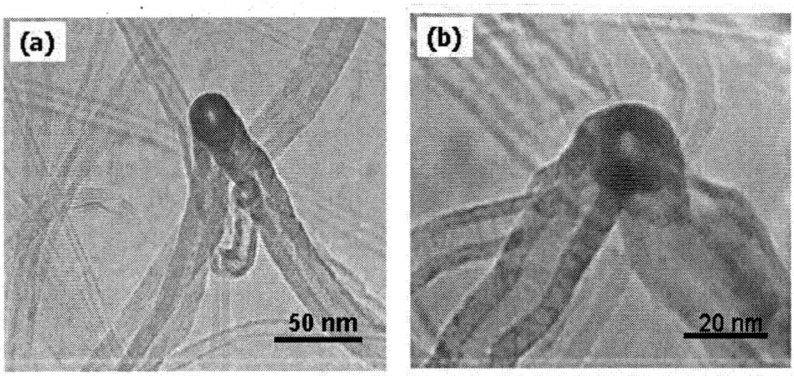

Fig. 23 TEM images of the fused junctions formed by a few CNTs joined together after the treatment of focused laser beam.

D.2 Investigation of the Growth Mechanisms of the CNTs

Our CNTs samples were synthesized via PECVD with the Fe nanoparticles as catalyst to assist the growth of CNTs. The growth mechanisms of CNTs are broadly classified into two categories- root-growth and tip-growth mechanisms. In the tip growth mechanism, the pentagon defects at the tips help in the further growth of the nanotubes beyond the initial stage. The Fe nanoparticle becomes encapsulated at the tip as the CNTs grew from the substrate. In the root growth mechanism, the reactants add on material to the roots of the nanotubes where the catalytic particle is anchored to the substrate. The CNT can be considered to be open ended at the root with carbon continuously added to the dangling bonds there. The rest of the CNT is then pushed up.

To investigate the mechanism of growth, it is common to take TEM image of the tip of the CNTs and to study if the tips consist of any inclusion. Fig. 23 shows TEM images of as grown CNTs synthesized using the PECVD system. On the one hand, we found tip inclusion (Fig. 23(a)) while no tip inclusion was found in another (Fig. 23(b)). As a result, it is not easy to draw any conclusion on the mechanism of growth.

Our laser-pruning method was useful in providing indirect evidence in determining the mechanism of growth of CNTs. The following procedure was adopted. Patterns were first cut on as grown well-aligned CNTs near the edges of the sample. The sample was then subjected to similar conditions for the growth of an additional layer of CNTs [23]. This will help to differentiate conclusively the sequence of the CNTs growth. Fig. 25 shows the result after this process.

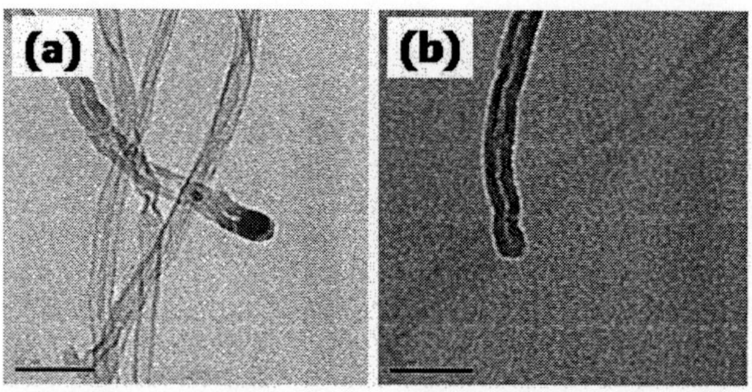

Fig. 24 TEM image of the tip of CNTs synthesized in this work. Scale bars: 50 nm

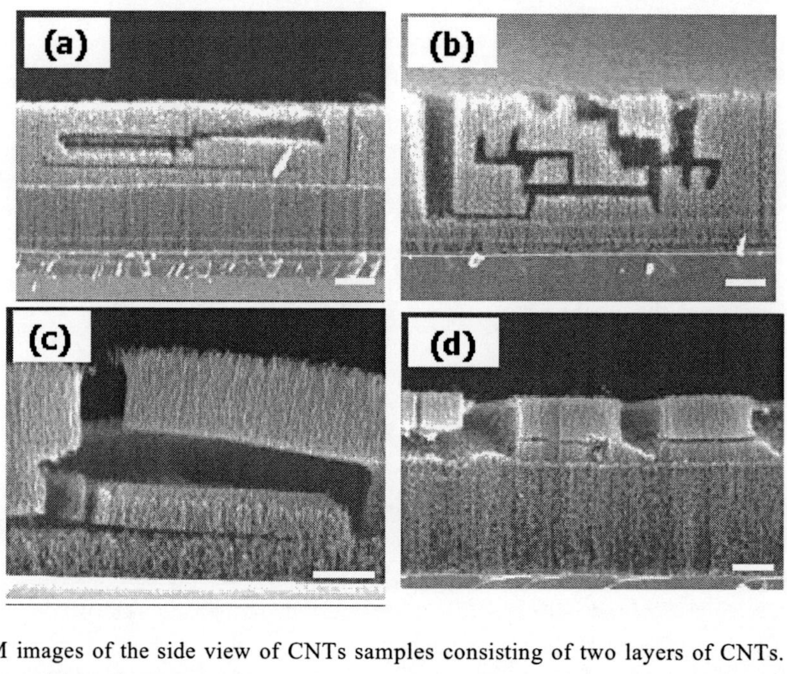

Fig. 25 SEM images of the side view of CNTs samples consisting of two layers of CNTs. Scale bars: (a,b,d) 10 μm and (c) = 5 μm.

As we have two recipes for the synthesis of CNTs (see section B), we can have four combinations of CNTs growth sequence. In all our samples, the first layer was grown with a growth time of 1200s (or 600s) in the first run and then for 1200s in the subsequent run. Fig. 25 shows SEM images of the side-view of the resultant samples. Fig. 25 (a) corresponds to sample where the CNTs were grown from a gas mixture of C_2H_2 and H_2 first followed by another layer of similar growth using the same recipe. Fig. 25 (b) corresponds to sample where the CNTs were first grown from the C_2H_2/H_2 gas mixture and followed by a second layer growth from C_2H_2/ NH_3 gas mixture. Fig 25 (c) corresponds to sample where the both layers of CNTs were grown with C_2H_2/ NH_3 gas mixture. Fig. 25(d) corresponds to sample where the first layer of CNTs was grown from C_2H_2/NH_3 gas mixture while the second layer was grown from C_2H_2/H_2 gas mixture.

The original patterns were lifted by continued growth of CNTs at the base of the original CNTs. A boundary that separated the 2 stages of growth was clearly visible and the patterns remained similar after re-growth. There was no observable further growth of CNTs at the tips during the second stage of growth. In the four different combinations of growth and re-growth employed, the initially grown CNTs are lifted up in the re-growth process. This points to a predominantly root-growth mechanism in the PECVD growth of CNTs in our samples.

In almost all instances of growth, it is noticed that the C_2H_2/H_2 recipe gives longer CNTs growth than the C_2H_2/NH_3 recipe for the same period of growth time. Although both hydrogen and ammonia are both reducing agents, one possible reason for the difference might be that diffusion of NH_3 molecules and the resulting plasma fragments are more difficult or slower than H_2 molecules and the resulting plasma fragments. This might have imposed a limiting factor on the iron nanoparticles' catalytic activity in the case of the C_2H_2/NH_3 recipe. Another visible observation is that the second layer of growth is generally shorter than the initially grown layer, given the same conditions and growth time. This is especially so in the case of growth using the NH_3 recipe as described above.

D.3 Laser-Induced Mechanical Motion of the CNTs

Aside from static CNTs structure, movable flaps of carbon nanotubes could also be constructed using this technique. By focusing a 29 mW laser beam normally onto a sample mounted in the face-up manner, parallel flaps of freestanding aligned nanotubes can be created. The CNTs flaps created can have a narrower base, compared to the top of the flaps. This is done by methods already described above. Fig. 26(a) shows the top view of a CNTs flap created by laser pruning using a 50x and long working distance lens. When a laser beam was focused at a spot between the flap and the wall with the sample mounted in the face-up position, the flap was found to deflect away from the focused spot as shown in Fig. 26(b). When the laser was blocked, the flap was found to return to its original positions promptly due to the elastic nature of the CNTs. Subsequently, as shown in Fig. 26(c), the long flap was broken into 2 portions by the focused laser beam. Similarly, the flaps deflected away from the laser spot when the laser light was present. However, freeing the original flap of one of the constraints imposed by the joined flap allows deflections of larger magnitudes to be observed. Videoclips of the laser induced deflection in the CNTs flaps can be found at the following website [25]. This technique can be extrapolated to configurations of more flaps with more laser spots controlling them.

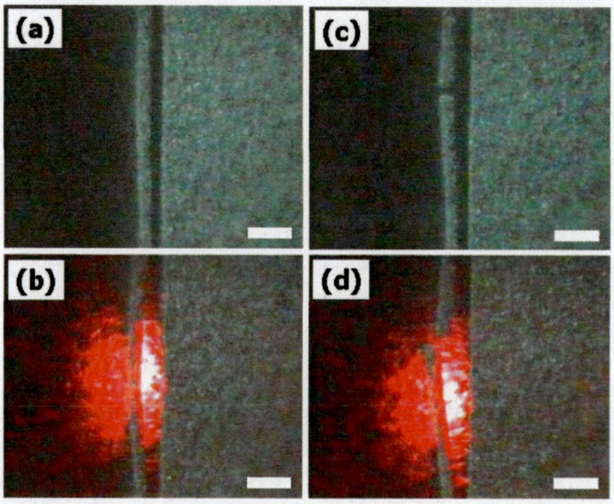

Fig. 26 (a,c) Optical Micrographs of flaps of CNTs fabricated via laser pruning. (b,d) Upon the illumination of the laser beam in the gap between the flaps and the main CNTs forest, the flaps were found to deflect to the left as shown. Scale bars: 20 μm. Videoclips of the process of laser mechanical deflection can be found at the following website [25].

The mechanism for the deflection of the flap by optical means is still under investigation although some possibilities have been proposed in similar studies involving nanotube bundles [26]. These possibilities include charge transfer in the CNTs as well as thermoelectric effects related to the electronic structure of the CNTs. More studies are still necessary to elucidate the underlying mechanisms.

D.4 CNTs Switches: Demonstration of the Feasibility of a Simple Switch with the Laser Induced Actuator

A simple optically activated switch constructed from vertically aligned CNTs flap was demonstrated. We made use of the circuit as shown in Fig. 27. The circuit consists of a multimeter (HP 34401A) and a voltage source (HP 6115A). One polarity of the voltage source was connected to an electrochemically etched tungsten tip and the opposite polarity of the source was then attached to the CNTs forest by means of silver tape. This forms a simple circuit where the circuit could be closed by the movement of the flap contacting the tungsten tip. The CNTs flap was constructed from vertically aligned CNTs by means of focused laser beams as in the methods described earlier. The tungsten tip was brought close to the flap but not touching it, as in Fig. 28(a). This corresponds to the "OFF" state. Focusing a beam of He-Ne (633nm) laser light by means of a 50X long working distance lens onto the CNTs forest near the flap caused the flap to deflect away from the laser spot as illustrated by Fig 28(b). This also allowed the flap to come into contact with the tungsten tip, thereby closing the circuit. This corresponds to the "ON" state, where a current can flow round the circuit.

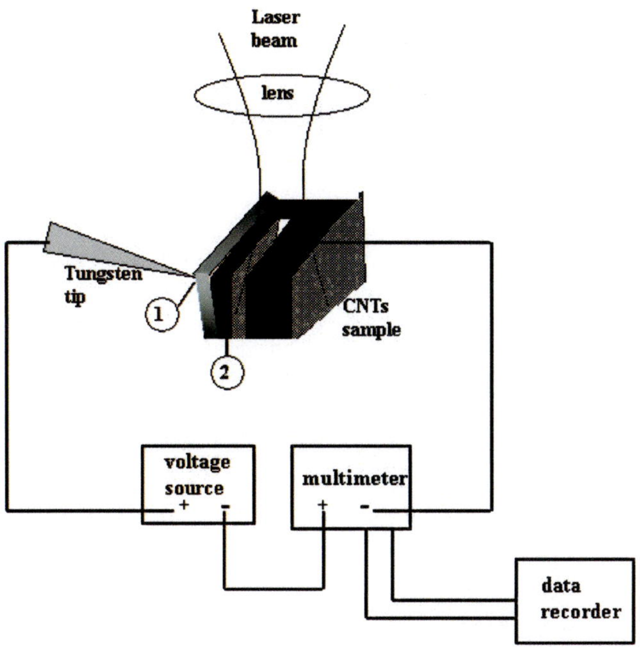

Fig. 27 Schematic of the setup for demonstration of an optically induced switch constructed of CNTs. Position 1 shows the position of the CNTs flap when the laser light was shone onto the system while position 2 shows the retracted position of the CNTs flap when the laser light was blocked.

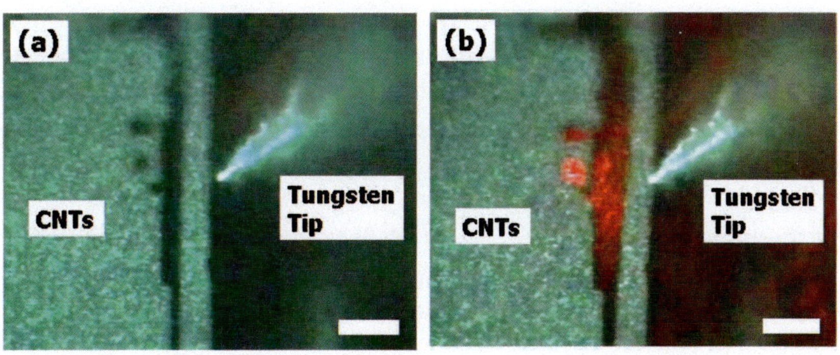

Fig. 28 (a) Optical Micrographs of a flap of CNTs near a sharp wire. (b) Upon the illumination of the laser beam in the gap between the flap and the main CNTs forest, the flap was found to deflect to the right and touch the sharp wire. Scale bars: 20 μm.

The "OFF" and "ON" states were then induced simply by externally blocking and unblocking the laser beam. This was achieved by using an optical chopper, which momentarily exposes the sample to laser light at regular intervals. At other times, the laser light was blocked. Fig.29 shows the current signals as recorded by the multi-meter. The spikes in the current represent times when the laser light was let in by the chopper to induce the deflection of the CNTs flap towards the tungsten tip. The periodicity in the optical chopper results in the current spikes at regular intervals. This optically actuated switch made of CNTs can thus provide a workable interface where an optical signal can be transformed into an electrical signal.

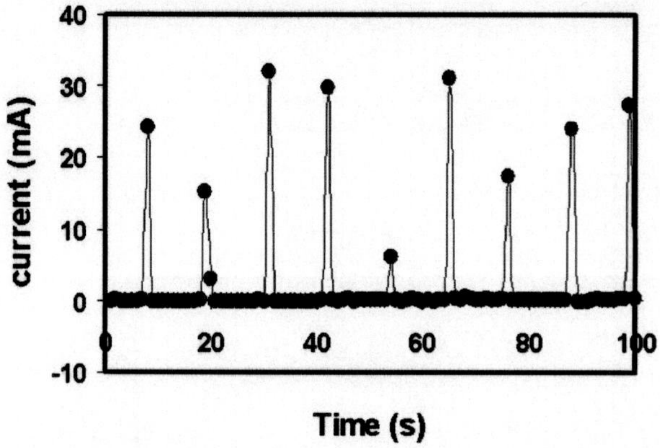

Fig. 29 Plot of electrical current across the system versus time.

E Experimental Technique II Laser-Pruning Using Pulse LAser

E.1 Experimental Setup

The focused laser beam technique developed and detailed thus far is well suited for the fabrication of unique 3D structures made of CNTs. However, since only a single focused laser beam was used, the process of fabrication is rather slow. It would be worthwhile to develop a complementary technique if patterns are desired in large extended areas of the CNTs forest. This second technique is built on a second laser-pruning technique that we have reported earlier in reference [24]. Using similar ideas of laser pruning of CNTs, we made use of a laser beam with large beam dimension to trim the CNTs forest. To define the required pattern onto the CNTs forest, we made used of a patterned copper grid as a mask.

Fig. 30 shows the schematics of the experimental arrangements used in this simple technique. The CNTs sample was first secured on a glass substrate and a copper grid was placed directly above the CNTs with spacers of thickness such that the copper grid is just above the CNTs tips. Copper grids consisting of regular arrays of both square holes or of a honeycomb pattern were utilized [27]. The sample was then exposed to a pulse (or a few pulses) of ultraviolet laser beam. The copper grid mask shielded the CNTs from the intense laser beam while the exposed CNTs were removed. As a result, we are able to generate patterns on the CNTs arrays defined by the copper grid masks.

In this work, we used a Lambda Physik Excimer Pulsed Laser (COMPex 201) commonly used for Pulsed Laser Deposition of materials [28]. The laser was filled with KrF gas and emitted UV light with a wavelength of 248 nm. It emits parallel laser beam with a broad beam dimension of 24 x 12 mm^2. Typically, we employed a laser energy density of about $1.7 mJ/mm^2$. Alternatively, we can position a converging lens in the optical path to create a slightly converging beam and thus increase its energy density for laser trimming of CNTs. In this experiment, we have exposed the same sample to different number of laser pulses and found that this enabled us to control the dimension of the features created in the CNTs forest. The damage on the grid was also found to be minimum and it could be re-used. This technique is fast, inherently parallel and is able to produce a large area of uniform patterned array of CNTs.

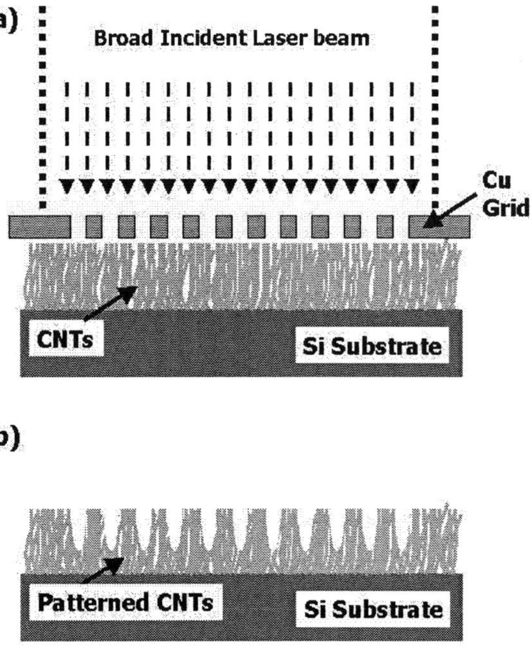

Fig. 30 Schematic of the second laser pruning method. (a) Illustration of broad laser beam incident onto CNTs forest masked with Copper gird. (b) Illustration of the resultant patterned CNTs.

E.2 Extended Areas of Patterned CNTs Arrays

Fig. 31(a) shows an SEM image of the copper grid consisting of a square array of holes used in this work. The dimension of each hole is 8 μm x 8 μm. Fig. 31(b) shows an SEM image of the array of holes generated on the CNTs covered substrate. The masked sample was exposed to 4 pulses of the laser beam and a square array of small holes with an average size of about 2 μm each was created. This was due to the converging nature of the laser beam employed where a larger fraction of the energy was concentrated at the center of each hole in the copper grid.

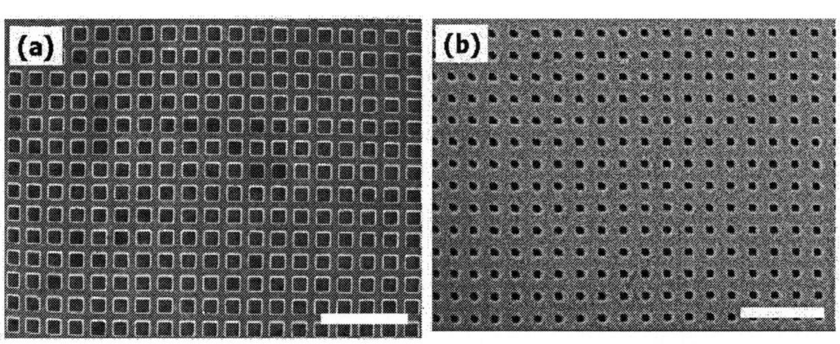

Fig. 31 (a) SEM image of copper grid used in this work with regular array of square holes. (b) Patterned-CNTs on the Silicon substrate after exposure to 4 laser pulses. A converging laser beam had been employed. Scale bars: 50 μm.

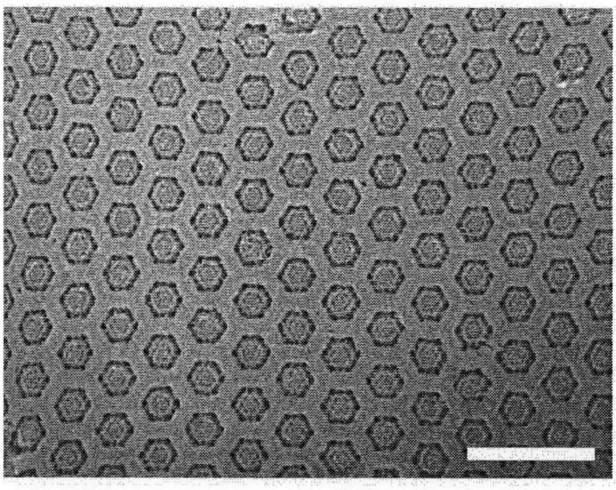

Fig. 32 SEM image of patterned CNTs created by masking it with a copper grid with a regular array of hexagons. Scale bar = 100 μm.

Fig. 32 shows another SEM image of the patterned CNTs sample fabricated using a different copper grid consisting of a regular array of hexagons. For this sample, only a single pulse was used. Again, the laser beam was slightly converging. The results indicate that one can make use of this technique to create an extended area of regular patterns on the CNTs samples. At the moment, the size of the patterned area was limited by the size of the copper grid. The size of the laser spot used in this work was about 140 mm^2 for a converging beam and 288 mm^2 for a parallel beam. Thus, as long as the laser beam intensity remains relatively uniform, one can create a patterned area as large as the laser beam size if a mask of comparable size is used.

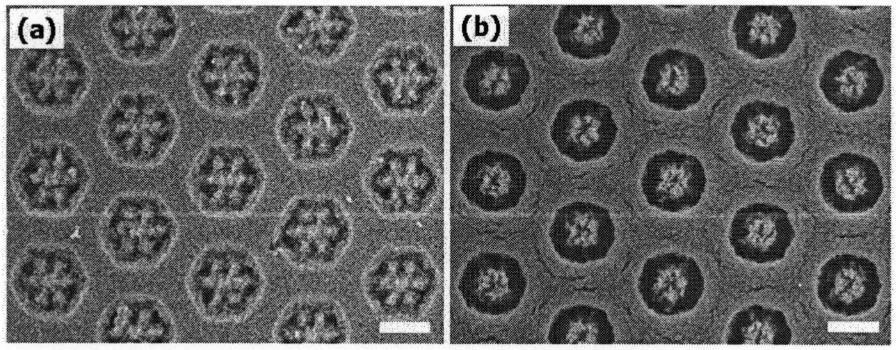

Fig. 33 SEM images of (a) patterned-CNTs after exposing to 1 pulse and (b) 2 pulses of laser. Scale bars: 20 μm.

In addition to the pattern defined by the copper grid, the diffraction patterns formed as a result of the laser passing through the copper grid were also capable of producing unique and interesting patterns on the CNTs sample. Fig. 33 shows SEM images of regular arrays of holes generated on the CNTs covered substrate after (a) 1 and (b) 2 laser pulses. Besides the pattern defined by the copper grid, additional patterns were created as a result of the interference pattern generated when the laser beam is incident upon the grid.

E.3 Laser Pruning of Carbon Nanotubes for Field Emission Devices

Carbon nanotubes has been identified as one of the best field emitters among all carbon based material discovered thus far. The high aspect ratio of the nanotubes, small diameters and long lengths, helps to generate a large electric field at the tip at low applied voltage. Thus, electron emission can occur in carbon nanotubes at low applied electric field. Aligned carbon nanotubes grown on silicon has been identified by to be good field emission array that can be easily incorporated into the standard silicon technology [17]. However, studies have shown that the proximity of the tubes can critically affect field emission properties of the nanotubes. This is due to the field-screening effect arising from the close proximity of the neighboring nanotubes [29]. Different techniques have been employed to alter the density of the carbon nanotubes, thus enhancing the effectiveness of the nanotubes as field emitters [30,31,32]. In this work, it was found that by exposing the CNTs sample to pulsed laser, the field emission performance of the CNTs sample could be enhanced. In this section, the preliminary results of the studies are presented.

Fig. 34 (a) SEM image (angled-view) of CNTs sample consisting of unexposed (I) and laser treated (II) region. (b,c) Close-up views of (b) region (I) and (c) region (II). (d) Fluorescent glow observed from the sample during field emission. Scale bars: (a-b) 10 μm, (c) 1 μm.

In this experiment, we divided the CNTs sample into three separate regions. Each region was exposed to a different number of laser pulses using the same Excimer Pulsed Laser as described earlier. Fig. 34(a) shows a SEM image of the boundary between an unexposed region and a region exposed to four laser pulses. In close-up views, Fig. 34(b,c) clearly show the distinct differences in the morphology and density of CNTs before and after the CNTs forest was exposed to the pulsed lasers. The laser caused the aligned nanotubes to adopt a more random and less dense nature. The sample was then placed in a typical Field Emission measurement unit. The anode was an Indium-Tin-Oxide coated glass further coated with fluorescent materials. Fig. 34(d) shows the remarkable difference in the intensity of the fluorescent glow for the different regions under an applied field of 1000 V. The as grown (unexposed) region did not produce any fluorescent glow. While the region exposed to the most number of pulses has the most intense florescence. Since, our pulsed laser technique can also define pattern of our desire, we can easily use it to control the density of the nanotubes at region desired for field emission. Thus, this technique will be a useful tool in controlling the density of the CNTs to improve the field emission performance of the large nanotubes array.

F Concluding Remarks

We demonstrate two simple techniques that allow for the pruning of nanotubes for a variety of purposes. The first technique allows for localized pruning of CNTs to achieve unique structures of both static and movable nature. The movable structures are generally CNTs flaps that can be actuated by laser light. This laser light activated movement is potentially useful for a variety of opto-electrical and opto-mechanical devices.

The second technique is a high throughput method capable of generating large arrays of patterned carbon nanotubes structures, useful for fabricating field emission display devices with interesting patterns. One possible usage of this technique is to reduce the density of CNTs on the substrate in instances where a medium CNTs density is more desirable for improved field emission from the sample [32].

Despite their individual strengths, both methods have the advantage of producing CNTs structures of different heights on the same sample. They can thus be viewed as techniques complementary to each other. By employing one or both techniques for the pruning of CNTs, a variety of interesting and useful structures can be envisioned.

Acknowledgement

Sow CH acknowledges the support of National University Singapore Nanoscience and Nanotechnology Initiative and Academic Research Fund.

References

[1] R. Saito, G. Dresselhaus, M. S. Dresselhaus, Physical Properties of Carbon nanotubes, 1998

[2] R. Martel, T. Schmidt, H.R. Shea, T. Hertel, Ph. Avouris, *Appl. Phys. Lett.*, **73**, 2447, 1998

[3] W.A.de Heer, A. Chatelaine, D. Ugarte, *Science*, **270**, 1179, 1995

[4] Dai H, Hafner J.H., A.G. Rinzler, D.J. Colbert, R.E. Smalley, *Nature*, **384**, 147, 1996

[5] R.H. Baughman, C.X. Cui, A.A. Zakhidov, Z. Iqbal, J.N. Barisci, G.M. Spinks, G.G. Wallace, A. Mazzoldi, D. De Rossi, A.G. Rinzler, O. Jaschinski, S. Roth, M. Kertesz, *Science* **284** (5418), 1340, 1999

[6] Li J, Papadopoulos C, Xu JM, Moskovits M, *Appl. Phy. Lett.* **75** (3), 367, 1999

[7] A C Dillon, K M Jones, T A Bekkendahl, C H Kiang, D S Bethune and M J Heben, *Nature*, **386**, 377, 1997

[8] de Heer WA, Chatelain A, Ugarte D, *Science* **270** (5239): 1179, 1995

[9] D. Sarangi, I. Arfaoui, J. M. Bonard *Physica B Condensed Matter*, **323** (1-4) (165-167), 2002

[10] Takashi Ikuno, Tetsuro Yamamoto, Motoki Kamizono, Syunji Takahashi, Hiroshi Furuta, Shin-ichi Honda, Shigeharu Ohkura, Mitsuhiro Katayama, Takashi Hirao and Kenjiro Oura, *Physica B* **323**, 171-173 2003

[11] Q. H. Wang, A. A. Setlur, J. M. Lauerhaas, J. Y. Dai, E. W. Seelig, R. P. H. Chang *Appl Physics Lett.* **72** (22), 2912-2913, 1998

[12] Jing Kong, Nathan R. Franklin, Chongwu Zhou, Micheal G. Chaplin, Shu Peng, Kyeongjae Cho and Hongjie Dai, *Science* **287**, 622, 2000

[13] G. Z. Yue, Q. Qiu, Bo Gao, Y. Cheng, J.Zhang, H. Shimoda, S. Chang, J. P. Lu, and O. Zhuo., *Applied Phyiscs letter* **81**, 355-357 2002

[14] Kenneth K. S. Lau, Jose Bico, Kenneth B. K. Teo, Manish Chhowalla, Gehan A. J. Amaratunga, William I. Milne, Gareth H. Mckinley, Karen K. Gleason, *Nanoletters*, **3**(12), 1701-1705, 2003

[15] K. C. Chin, A. Gohel, H. I. Elim, W. Ji, G. L. Chong, K. Y. Lim, C. H. Sow, A. T. S. Wee *Chemical Physics Letters* **383**, 72, 2004; A. Bianco and M. *Prato Advanced Materials* **15** (20), 1765, 2003

[16] Ali Javey, Qian Wang, Ant Ural, Yiming Li and Hongjie Dai *NanoLetters*, **2**(9), 929-932, 2002

[17] Fan S, Chapline MG, Franklin N R, Tombler T W, Cassell A M and Dai H, *Science,* **283**, 512, 1999

[18] Z. Y. Zhang, B. Q. Wei, g. Ramanath and P. M. Ajayan *Applied Physics Letters* **3764**-3766 77 (23) 2000

[19] K.B.K. Teo, M. Chhowalla, G.A.J. Amaratunga, W.I. Milne, D.G. Hasko, G. Pirio, P. Legagneux, F. Wyczisk, D. Pribat, *Appl. Phys. Lett.*, **79**, 1534, 2001

[20] B.Q. Wei, R. Vajtai, Y. Jung, J. Ward, R. Zhang, G. Ramanath, P.M. Ajayan, *Nature*, **416**, 495, 2002

[21] Z. P. Huang, D. L. Carnahan, J. Rybczynski , M Giersig, M. Sennett, D. Z. Wang, J .G. Wen, K. Kempa, Z. F. Ren *Applied Physics Letters* **82** (3) 460, 2003

[22] Yoshikazu Homma, Yoshihiro Kobayashi, Toshio Ogino, Takayuki Yamashita *Applied Physics Letters* **81**[12], 2261, 2002

[23] K. Y. Lim, Chorng Haur Sow, Jianyi Lin, Fook Chiong Cheong, Ze Xiang Shen, John Thiam Leong Thong, Kok Chung Chin, and Andrew Thye Shen Wee *Advanced Materials* **15**(4), 300-304, 2003

[24] Fook Chiong Cheong, Kim Yong Lim, Chorng Haur Sow, Jianyi Lim and Chong Kim Ong *Nanotechnology*, **14**, 433-437, 2003

[25] Video clips of the laser pruning of CNTs and laser induced motion in the CNTs flaps can be found at the following website: http://www.physics.nus.edu.sg/~physowch/CNTNova/CNTNova.html

[26] Y. Zhang, S. Iijima, *Physical Review Letters*, **82**, 3472, 1999

[27] AGAR SCIENTIFIC LIMITED, 66A Cambridge Road, Stansted, Essex, CM24 8DA, England.

[28] Xu S Y, Low B L, Ong C K and Zhang X., *Journal of Materials Science Letters* **16** 429, 1997

[29] Suh JS, Jeong KS, Lee JS, Han IT *Applied Physics Letters* **80** (13): 2392-2394, 2002

[30] Bae EJ, Choi WB, Jeong KS, Chu JU, Park GS, Song S, Yoo IK *Advanced Materials* **14** (4), 277, 2002

[31] Semet V, Binh V T, Vincent P, Guillot D., Teo K B K, Chhowalla M, Amaratunga G A J, Milne W I, Legagneux P. Pribat D *Applied Physics Letters* **81** (2), 343-345, 2002

[32] Nilsson L, Groening O, Emmenegger C, Kuettel O, Schaller E, Schlapbach L, Kind H, Bonard J M and Kern K, *Appl. Phys. Lett.* **76** 2071, 2000

In: Focus on Nanotube Research
Editor: Delores A. Martin, pp. 97-126

ISBN: 1-59454-851-X

Chapter 5

CARBON NANOTUBE-POLYMER NANOCOMPOSITES: PRINCIPLES AND APPLICATIONS

Carlos Velasco-Santos[†, ‡, §]***, Ana L. Martínez-Hernández***[†, ‡, §]
and Víctor. M. Castaño[†]
Centro de Física Aplicada y Tecnología Avanzada,
Universidad Nacional Autónoma de México A.P. 1-1010, Santiago de Querétaro, Querétaro 76000, MÉXICO, Department of Materials Science, University of North Texas, Denton, TX 76203-5310, USA, Departamento de Materiales, Departamento de Mecatrónica, Instituto Tecnológico de Querétaro, Av. Tecnológico, Col. Centro, Santiago de Querétaro, Querétaro 76000 MÉXICO
[†] Universidad Nacional Autónoma de México
‡ University of North Texas
[§] Tecnológico de Querétaro

Abstract

Carbon nanotubes are undoubted one of the most interesting materials in the last twenty years. The outstanding properties that these materials possess have opened a new interesting research area in nanoscience and nanotechnology, in addition, new properties have been found and possible applications are suggested for these materials frequently. Nowadays the research in nanotubes covers different fields focus to understand and applied the interesting properties of these nanomaterials, and other part have been directed to basic science with the target to modify structure and surface of carbon nanotubes. This later has allowed diversify their properties and possible uses, opening new possibilities in chemistry, physics, engineering, medicine and materials science. One of the most mentioned applications for these carbon materials is their incorporation as reinforcement in polymer matrices in order to produce new nanocomposites, taking advantage of the amazing structural, mechanical, electrical, chemical and thermal properties that carbon nanotube possess. In this context carbon nanotube nanocomposites have been developed by different research groups with the aim to produce new ultra-strong and light materials, in addition electrical conductivity and thermal properties have been studied for this kind of nanocomposites. Nevertheless, a new challenge that allows making a new age of multifunctional composite materials with these nanometric forms is starting and therefore the study of new properties in these nanocomposites is increasing significantly by scientific groups in the last five years. In this chapter we present different results of mechanical, thermal and electrical properties in carbon nanotube polymer composites, with special emphasis in novel anisotropic tribological and conductive properties that have been found recently in these nanocomposites. Chemical functionalization with focus to improve compatibility between these nanoforms and polymers is also discussed. In addition are presented the results obtained with polymer nanocomposites developments with modified nanotubes. The relevance of understand the behavior on the interface between nanotubes and polymer matrices to control and improve different properties and this way reach outstanding and novel multifunctional applications with this materials is discussed continuously.

1 Introduction

Some years ago nanotechnology applications seemed to be very distant. However with discover of new nanomaterials with impressive properties never seen before, nanoscience research got to increase notably and therefore nanotechnology applications and products have started to appear recently. Nevertheless nanotechnology applications are beginning to come out and more research is needed to reach new novel results and uses for these kinds of materials. In this context development of nanocomposites is a research area growing very fast, inasmuch as has been found that few loads of nanoreinforcement can improve significantly different properties in structural matrices when a composite is developed. One of the most notable nanomaterials in this field is indubitable carbon nanotubes (CNs), discovered by Iijima in 1991[1]. These materials have been proposed for diverse interesting nanotechnology applications and the development of nanocomposites is one of them.

Carbon nanotubes have outstanding mechanical properties, such as extremely high Young's modulus, stiffness and flexibility [2-6], and the electronic properties of nanotubes allow to be applied as semiconducting and metallic conductive nanomaterials [7]. As well these nanostructures have shown high thermal stability [8].

All above mentioned, together with their nearly perfect structure, small diameter, high surface area and high aspect ratio, offer an amazing structure with unique properties useful to reinforce polymer matrices. This represents attractive possibilities for developing new ultra-light and strong composite materials and creates alternatives to extend the functions of organic materials when a composite is developed due to their thermal and electrical properties. Thus carbon nanotubes properties exceeds most common materials used as reinforcement, and the properties diversity than present these nanoforms provide the possibility to develop new multifunctional composites. Even thought interesting results have been reached in metal and ceramics matrices reinforced with nanotubes [9-12]. At the moment, most relevant results have been found when polymer matrices have been modified with carbon nanotubes. In spite of this, research in carbon nanotube polymer nanocomposites (CNPN) is a relatively new field and many aspects at interface level to produce diverse properties in these materials are in current research.

Different reports have studied the incorporation of CNs in polymers seeking a synergic effect in the composite, caused by the good interaction on interface level that can be distinguished in diverse properties of the material. Nonetheless results have been too dissimilar in some cases, as a result of the fact that several kinds of CNs have been used. At the present time there exist different types of nanotubes such as single-walled nanotubes (SWNTs), double-walled nanotubes (DWNTs), multiwalled nanotubes (MWNTs) and their variants with helix and bamboo forms.[13] These materials can be produced from arc discharge[14-15] , pyrolysis of hydrocarbons over catalysts[16] , laser vaporization of graphite[17], electrolysis of metal salts with graphite electrodes[18] to hydrothermal methods[19] , and recently one type of nanotubes have been found in natural environment [20]..

SWNTs and MWNTs are in fact the nanostructures with more studies around them, inasmuch as these nanostructures were discovered first of all. Thus, research in carbon nanotube nanocomposites have been developed in the majority with SWNTs, and MWNTs, This have represented an important variable, inasmuch as the difference on the number of walls produce variation on the diameter causing different interesting characteristics on the composites. In addition the production approach employed to synthesize CNs is not always the same, and cause different graphitization degrees and diverse properties in the material. Also various polymer matrices have been used with different processing methods. Besides of all previously mentioned, recent interesting researches have employed functionalized carbon nanotubes (f-CNs) in order to improve interface between nanotubes and polymer matrix. On the whole these and other factors have been handled in different researches showing that different parameters influence the dispersion and interface, playing essential role in the composite behavior. So, even though, some results in carbon nanotube polymer composites have been opposite, the majority of these works are the base to develop a new age of strong, conductive and smart materials in base of CNs and polymers. Now there exists enough evidence to believe that it's possible to form multifunctional nanocomposites, inasmuch as new novel properties in nanocomposites have been found.

In multifunctional nanocomposites not only one mechanical or electrical property will be reached, but different relevant properties can be obtained in the same material. This could be very useful to create novel nanocomposites and smart materials for a diversity of applications. Nevertheless research in carbon nanotube polymer nanocomposites is practically beginning and it's important to known which are the achievements reached in this matter.

Thus, we present the background in mechanical, thermal and electrical properties in carbon nanotube polymer nanocomposites, in addition are discussed the influence of chemical modification in carbon surface and the relevance that this kind of functionalization have had in the mechanical thermal and electrical properties in nanocomposites. In addition novel anisotropic behavior found in these types of composites materials is presented. Finally are discussed the possible applications of carbon nanotube polymer nanocomposites in base of the total results presented in this chapter.

2 Load Transfer and Alignment

As we mentioned before, interesting results in mechanical, thermal and conductive properties on carbon nanotube polymer nanocomposites (CNPN) have been reported in the last six o seven years. Nevertheless, the pioneers' researches that involved the incorporation of nanotubes in polymer matrices had as aim to study the dispersion, aligned, deformation and possible load transfer that these materials present in polymer materials. These topics are still considered research purposes; inasmuch as new physical and chemical modifications and additives are proposed constantly in order to improve these parameters. The first paper in carbon nanotube polymer nanocomposites was published eleven years ago. In this study

Ajayan et al.[21] aligned CNs in epoxy, then, they cut small slides in order to get a partial aligned segment. Aligned carbon nanotubes in polymer matrices can be consider as a designed arrangement in some cases, inasmuch as this would produce a control of the properties in the direction of aligned carbon nanotube in the nanocomposites. However this first evidence of aligning CNs in polymer matrix just applies to diminutive areas of these nanocomposites.

Jin et al.[22] published alignment of CNs in polymer matrix by mechanical stretching. They employed MWNTs synthesized by arc-discharge method and polyhydroxyaminoether as matrix. Thermoplastic polymer was dissolved in chloroform and then molded in teflon. The results showed a partial alignment that depends on the stretching ratios and on the carbon nanotubes fraction. Alignment by stretching is not new approach in fiber reinforced polymer composites; however few results was known with respect to carbon nanotube and these contributions presented the first results of nanotube behavior as reinforce of polymers.

In other word alignment of SWNTs was reached by a combination of solvent casting and melt processing methods, there the CNs were dispersed in Polymethyl-methacrylate (PMMA) using as solvent dimethylformamide. The arrangement in these composites allowed a higher conductivity behavior in flow direction. Also Elastic modulus (E) and yield strength (Y) were measured in melt spun fibers, both E and Y increase with nanotubes content and draw ratio. Although these increments are not uniform, in both parameters higher values are reached with 5 wt% and 8 wt% of SWNTs [23].

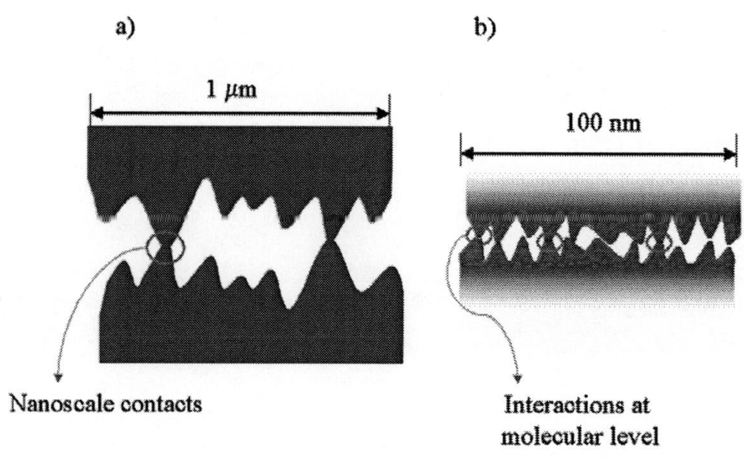

Figure. 1 Interactions at interface level, a) micrometric scale, b) nanometric scale.

Wagner et al. developed an interesting study where stress transfer ability was probed in carbon nanotube polymer nanocomposite. In this report was observed the fragmentation of MWNTs under tensile stresses in thin films. The results show that the effectiveness to transfer stress in this composite is at least one order of magnitude larger than in conventional fiber composites[24]. This kind of reports was one of the first important contributions that gave an idea about the significance of structure and size of carbon nanotubes at interface level. Inasmuch as in theory the interactions between nanostructures in tubular form and polymer matrices should be better with respect to conventional fiber with diameters in millimeters or micros. Nevertheless nanostructures can to form bundles or clusters and to show different behavior in the interface. Thus, to know which interactions at nanometric can achieve efficiently to transfer load with carbon nanotubes was a starting point which occasioned different researches in this field. Figure 1. illustrate the differences between nanometric and microscopic contacts at interface level. Figure 2. show the damage produced in adjacent carbon nanotubes surrounded by polymer matrix.

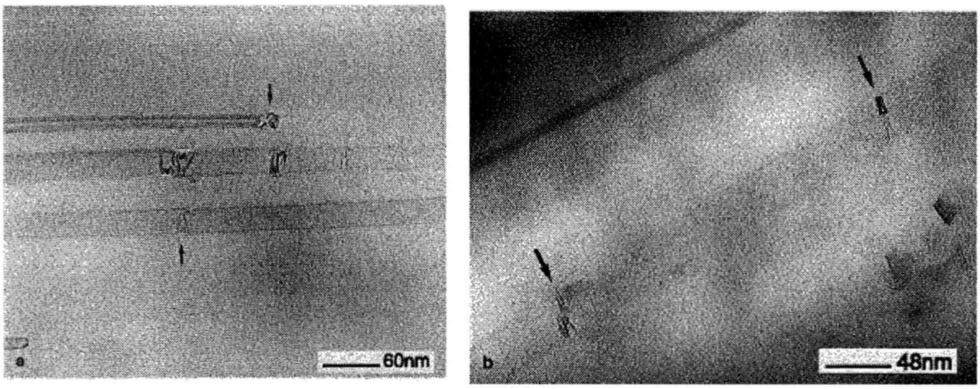

Figure. 2 Damage in nanotubes close to the polymer film surface (a) Transmission Electron Microscopy (TEM) image of nanotube fragmentation and the formation of a doublet of adjacent breaks (bottom arrow). A doublet consisting of a nanotube end and a nanotube break is also shown (top arrow). (b) TEM picture of two aligned multilayer carbon nanotubes in close contact, revealing the formation of two doublets of adjacent breaks. *Reprinted from the publication* [51] *, Copyright Elsevier 1999.*

A later report of Lourie and Wagner signs on that the polymer nanotube interface is not inert and there exist load transfer to the nanotubes via surrounding matrix. In this paper it was suggested that in spite of the huge difference in scale between traditional fiber composites and carbon nanotube nanocomposites, the fundamental concepts in continuum mechanics of conventional composites could have some degree of validity at nanometric level[25]. However, nowadays it's known that the application of a model to these types of materials depends of many factors and in the several cases, continuum mechanics of conventional composites does not apply in nanocomposites. Different works have been developed in the case of carbon nanotube polymer modeling. An example is the report published by Fisher et al [26-27]. There, they used a combined finite element and micromechanical approach to determine the influence of curvature in carbon nanotubes when they are incorporated in polymer matrices. In this work they assumed that CNs and matrix are isotropic materials with equal Poisson ratios. The results of this model suggest that waviness in nanotubes could be other important factor that may limit the modulus improvement compared with straight nanotubes. Figure 3. show the effective modulus behavior of carbon nanotubes polymer nanocomposites both experimental data and micromechanical predictions.

Other studies that have shown load transfer have been developed by deformation studies in transmission electron microscopy (TEM). In this research was used a polystyrene matrix containing MWNTs to form the nanocomposite. The studies were realized using TEM straining holder and film with a notch. The results show that nanotubes lie to align and bridge the crack when a tensile stress was applied to the composite, showing load transfer across the nanotube-polystyrene interface [28].

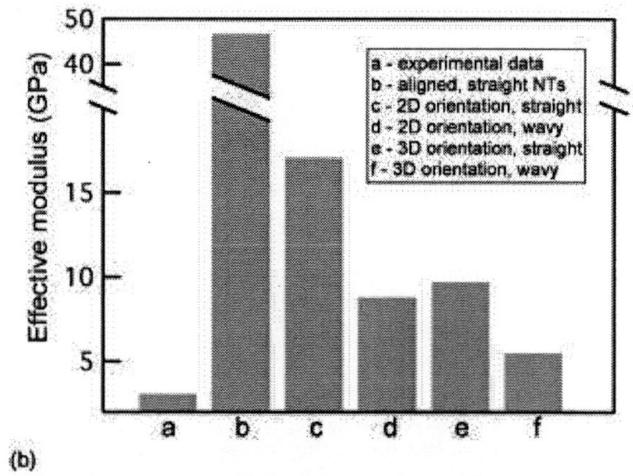

Figure 3. Experimental data (10 wt % MWNTs in polystyrene) and micromechanical predictions (assuming straight and wavy nanotubes) of nanotube reinforced polymer effective moduli with different CNs orientations. *Copyright AIP 2002.*[26]

Poly (methyl methacrylate) (PMMA) has been studied as polymer matrix of carbon nanotubes in different researches. Parameters such as CNs distribution and the possible modification of CNs when they are incorporated in polymer matrix have been analyzed in this polymer. Investigation developed with these aims shown that low concentrations of SWNTs obtained from arc-discharge method lie to give uniform distribution in PMMA polymer [29-30]. Nevertheless different results have been obtained with this polymer with respect to the CNs distribution and compatibility some of them are discussed later in this chapter.

3 Mechanical Properties in Carbon Nanotube Polymer Nanocomposites

Carbon nanotube-polymer nanocomposites are in fact, research field in the whole world. Investigations have provided diverse results in this kind of composites, mainly with respect to the mechanical properties, showing that many factors play an important role in the synthesis and influence the interface affecting therefore carbon nanotube properties.

Different approaches have been used to produce carbon nanotube polymer nanocomposites (CNPN), some of the processes used are: melt blend [31], dissolve and casting [32], "*in situ* "polymerization[33] and extrusion[34] among others. Some interesting researches where different synthesis methods were used are presented next. Results show diverse mechanical and thermal properties which allow establishing some relevant conclusion in CNPN.

Schadler et al.[35] developed MWNTs-epoxy composite cured with triethylene tetraamine. They evaluated both moduli: tension and compression. The results show that compression modulus is higher than tensile modulus in this composite. The load transfer was evaluated through Raman spectroscopy, and it was found that strain in carbon bonds only shift significantly under compression. The authors sign on that all the walls in MWNTs only participate in compression, unlike when the composite is subject to tension stress, inasmuch as is proposed that in this case only take part the outer walls of carbon nanotubes.

Shaffer and Windle produced carbon nanotube-polyvinyl-alcohol nanocomposite with different loads of catalytically grown nanotubes, from 10 wt% to 60 wt%. Dynamic mechanical parameters were measured in these CNPN. The values of storage modulus (E') were incremented by more than two folds but with a huge quantity of CNs, which is not the most expected in this kind of composites. At the same time the onset thermal degradation was retarded with these composites, suggesting the application of CNs as modifiers.[36] In other study MWNTs-PMMA composites were developed and probed mechanically. Two types of MWNTs were incorporated to polymer matrix: as-synthesized and treated MWNTs (t-MWNTs are CNs ground in a high rate ball mill for 20 min and then boiled for 0.5 h in nitric acid), the method used to form the composites was polymerization using azoiso-butyronitrile (AIBN) as initiator. Satisfactory increments in the tensile strength and hardness were reached with 3 wt% and 5 wt% of t-MWNTs, however with more than these loads the tensile strength lies to down. The relevance of the first results is attributed for the authors to the good interaction between CNs and polymer matrix, this is due to the initiator could break the π bonds of CNs, and in this way it is possible to forms a chemically bonded interface. However many factors could take part in this success, since t-MWNTs were treated with acid nitric solution. This could cause a slight oxidation in some zones of the nanotubes, where the defects might be increased by the initiator action, producing functionalized nanotubes that would interact with the polymer chains. More research is needed to confirm this supposition and to understand the role that could play the polymeric initiator in CNs bonds and in the development of this type of nanocomposites [37].

As we mentioned before PMMA is one of the most studied matrix in the field of these types of composites. Cooper et al. evaluated mechanical properties of PMMA nanocomposites formed with MWNTs, SWNTs and nanofibrils. The results of these analyses present a tensile modulus almost insensitive to the presence of CNs; however impact properties are significantly improved by both MWNTs and SWNTs [38]. Tatro et al., in other recent study with PMMA and MWNTs, have evaluated thermal and electrical properties, Vickers microhardness and dynamical mechanical analysis in nanocomposites irradiated with a Cesium-137 source. The target was to determinate the effect that might have carbon nanotubes reinforcing polymer matrix when the sample is irradiated. The results show that MWNTs may improve radiation hardness of mechanical properties and glass transition temperature when a composite is formed. The effect is attributed to conjugated bonds in CNs which may absorb some part of the radiation energy, limiting in this way the damage of PMMA molecules. More aging studies with radiation and evaluation of the interface are proposed and currently developed by these authors [39].

Other recent report evaluates the dynamical mechanical properties of pure PMMA and nanocomposites formed with SWNTs and PMMA matrix at two different frequencies. The study was developed with the aim to understand the effect produced by the incorporation of the nanotubes on the ability of PMMA to store and dissipate the mechanical energy. SWNTs used in this research were produced by laser ablation method and the nanocomposites were prepared by "*in situ*" polymerization using a solution of dimethylformamide (DMF) and 2,2-azobisisobutyronitrile (AIBN) as initiator. α-relaxation, related with the onset of the glass transition temperature (Tg) and β-transition associated with hindered rotation of the side chain, were evaluated in PMMA and SWNTs–PMMA nanocomposites. Transitions were detected in both tangent delta curves and elastic modulus curves (E') at low and high temperatures. Results in two different frequencies show that PMMA and nanocomposites formed with slight quantity of SWNTs exhibit similar β-transition temperature and identical Tg's. However E' at low temperature (-150°C) in the nanocomposite with a tiny quantity of 0.014 wt % of SWNTs, exhibits an important increment, in comparison with E' of the pure PMMA, from ≈ 5.1 GPa in PMMA to 7.2 GPa for the nanocomposite, both evaluated with a frequency of 100 rad/s. This represents a significant rise considering the slight quantity of SWNTs that were used.

In addition the research involves different studies such as Termogravimetric Analysis (TGA) and Nuclear Magnetic Resonance (NMR) in order to evaluate possible changes in the structure of the polymer nanocomposite caused by the addition of SWNTs. Neither microstructure nor thermal degradation changes were localized in these nanocomposites in comparison of pure PMMA. This discards that these parameters could be liable for the change in the mechanical properties of the nanocomposites. Therefore it is signed on that the interface in these nanocomposites plays an important role; and the properties of the polymer are changed in the vicinity of the SWNTs by the cohesive interactions, caused by the large surface area of CNs [40].

As well this research involves the discussion and possible adaptation of the traditional composite theory in the case of CNs nanocomposites; however the authors in agree with other researches mention that nanotubes have a great deal of effect on the properties of the composites, larger than those expected for the traditional composite theory [40]. Therefore it is possible that new models will be needed in order to adjust and predict carbon nanotube nanocomposites properties, inasmuch as this nanomaterials and the interface produced with polymers are causing unexpected properties.

Dufresne et al. processed and characterized nanocomposites, which were developed with catalyzed MWNTs and poly(styrene-co-butyl acrylate) [41]. The composites were prepared by casting after stirring with different loads of CNs from 0 to 15 wt %. Mechanical properties of these composites were evaluated by dynamical mechanical analysis (DMA) and tensile test, finding increments in the tensile modulus with the content of carbon nanotube almost in all samples. However unexpected decrease in the modulus was observed in the sample with 7 wt% of CNs. Thus although a good distribution is reached in the majority of the samples, some areas with nanotubes appear like clusters causing unpredictable behavior in some samples, inasmuch as is difficult to know how many of these aggregates appear in the composite. This has been observed in other composites prepared with carbon nanotubes when the weight percent is higher than 5 or 7wt % [32] and explains the unexpected behavior in this study. However it is possible than not only this factor affects the development of these materials with relative high concentrations. Therefore it's needed to continue with the research within a similar context.

Wong et al. have analyzed two kinds of nanocomposites that show agglomerates, these cause similar effects than those mentioned before, given that mechanical behavior lies to down in some cases when the content of CNs was incremented. Although it was probed by field emission scanning electron microscopy (FESEM) and transmission electron microscopy (TEM) that CNs have good interaction and adhesion with the polymers, other regions show agglomerates that impede the adequate distribution and cause the decrease in the nanocomposite properties. In this research CNs reinforcing polystyrene rod and epoxy thin film were analyzed. In the case of CNs polystyrene composites, tensile strength increases from 22.1 MPa (for polystyrene) to 24.4 MPa (for the sample with 0.1 wt% of CNs), however when the content of CNs was incremented at 5 wt% the strength is diminished until 17.9 MPa. Besides the samples with 1 wt% and 2 wt% show lower values in tensile strength than the sample of pure polystyrene [42].

Also in this study, epoxy thin films were evaluated mechanically. Microscopy results of these samples revealed the two mentioned regions. One corresponds to the failure of the matrix but not on the CN-epoxy interface; nevertheless in other regions CNs agglomerates were observed showing a local poor dispersion. Authors sign on that this later behavior is the possible cause of the failure initiation. Similar results where some agglomerates affect mechanical properties mainly at relatively high concentration have been found in poly (methyl-ethyl methacrylate) [32]. Figure 4. shows the storage modulus (E') obtained by DMA (Dynamical Mechanical Analysis) for the samples which were produced by dissolution of copolymer . The most outstanding modulus is obtained when only 1 wt % CN without additives (sample 1) were used, increasing the modulus by more than 200% at 40°C, which represents one of the most relevant increment reached until now at room temperature with the incorporation of very low quantity of CNs [43].

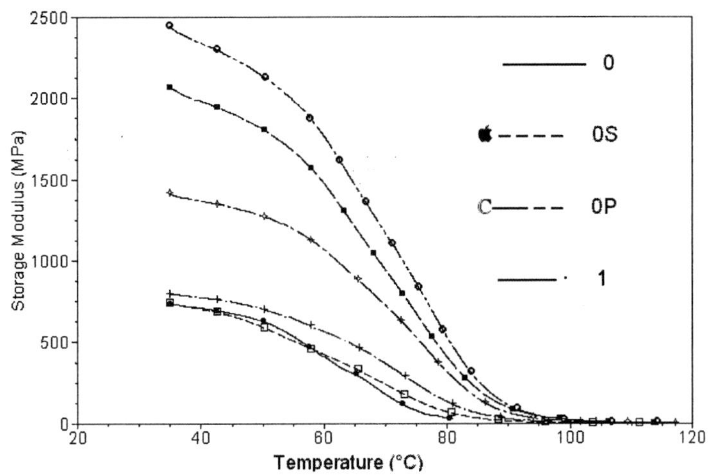

Figure 4. Storage Modulus of Carbon nanotubes films: a) samples: 0 (MEMA), 0S (MEMA - 1wt% surfactant), 0P (MEMA - 1wt% plasticizer), 1(MEMA – 1wt% CNTs), 1S (MEMA – 1wt% surfactant-1wt% CNTs), 1P (MEMA – 1wt& plasticizer- 1wt% CNTs); *Copyright IOP 2003* [32].

Figure 5 shows the results of storage modulus (E') obtained in DMA for the samples where an surfactant was used (0S, 1S, 5S, 7S, 10S), E' is increased in the samples with 1 wt % (1S) and 5 wt % of CN (5S) with respect to the sample with only 1 wt% of surfactant, nevertheless in the samples with 7 wt % (7S) and 10 wt % (10S), E' diminish with respect to the sample 5S and 1S. As we mentioned this behavior have been presented in other reports where more than 5 wt % of nanotubes were used, inasmuch as CNs form cluster zones which difficult the interaction at interface, avoiding that several nanotubes maintain contact with polymer matrix. Figure 6 show a Scanning Electron Microscopy image of an agglomerate found in these nanocomposites. Figure 7. shows comparable bundle found in the polystyrene nanocomposite developed in reference [42]. In these and other researches big quantities have became the reason of modulus or other properties lie to down, inasmuch as the produced agglomerates, avoid both reaching good dispersion and taking advantage in the best way of CNs properties, through an adequate load transfer.

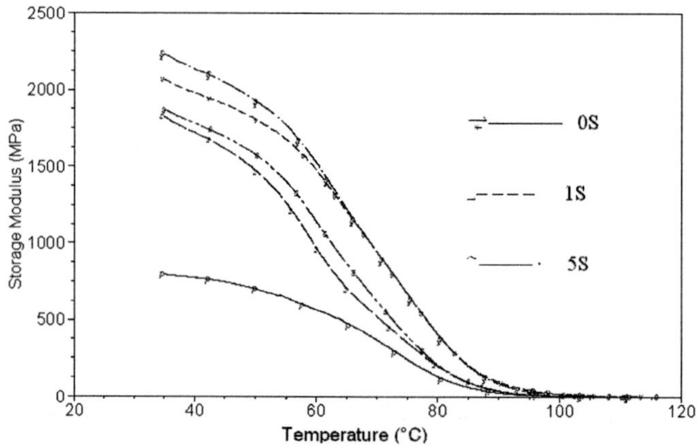

Figure 5. Storage Modulus of Carbon nanotubes films: 0S (MEMA - 1wt% surfactant), 1S (MEMA - 1wt% surfactant- 1wt% CNTs), 5S (MEMA -5wt% surfactant- 5 wt% CNTs), 7S (MEMA -7wt% surfactant- 7wt% CNTs), 10S (MEMA -10wt& surfactant- 10wt% CNTs). *Copyright IOP 2003* [32]

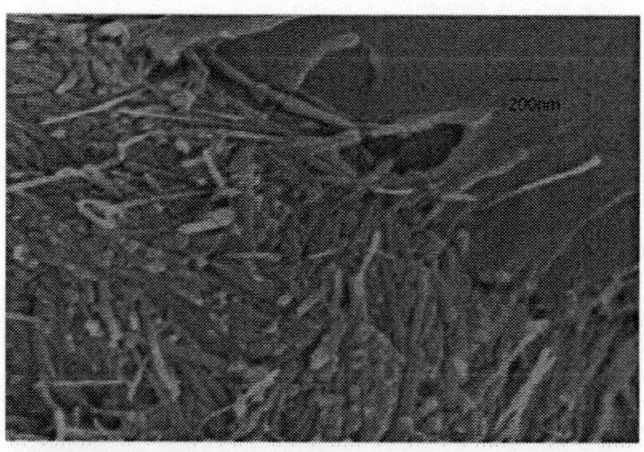

Figure 6. Scanning Electron Microscopyn (SEM) image of carbon nanotube composites showing a bundles of CNTs in polymer matrix. *Copyright IOP 2003* [32]

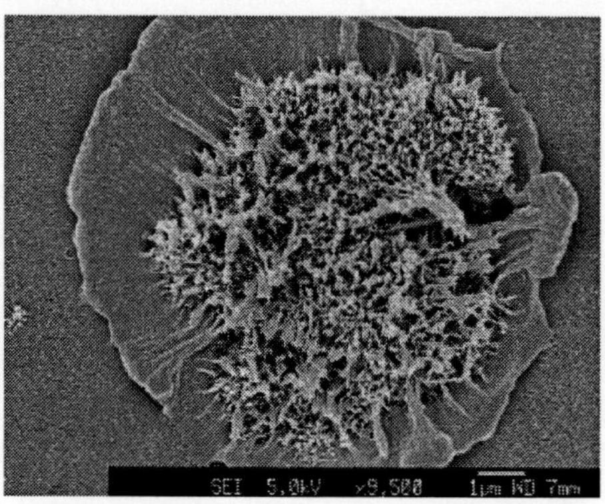

Figure 7. Field Emission SEM image of fracture surface of CNT-PS composite: a CNT agglomerate being pulled half way out from the Polystyrene matriz. *Copyright Elsevier 2003 [*62]

Wong et al. furthermore show an interesting analysis which involves molecular simulations of CNs pullout. Molecular models of SWNTs in polystyrene matrix and SWNTs in epoxy matrix were generated, and then total work required for fiber pullout was calculated. Figure 8 show a molecular model of CNs in polystyrene matrix. The results obtained for CN-epoxy are very similar than those obtained in experimental pullout for the same system. Authors conclude that the results of this simulation and the previous experimental studies suggest that the interfacial shear stress of CNs polymer systems is at least an order of magnitude higher than micro fiber reinforced composites and it is assumed that the surface of carbon nanotube is important to get an intimate contact at nanometric level between CNs and polymer matrix[42].

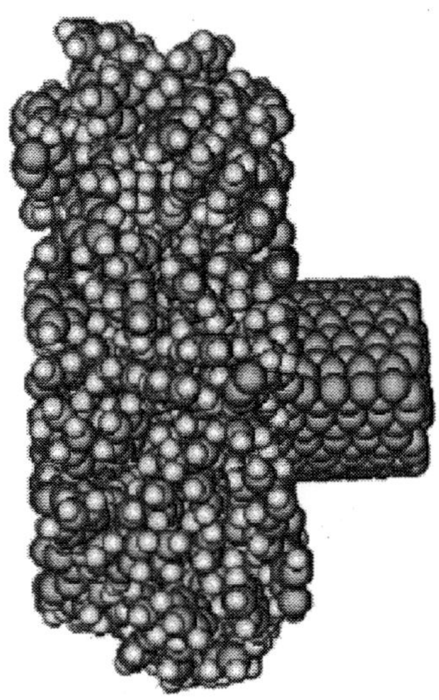

Figure 8. A molecular model of a carbon nanotube in polystyrene matrix. *Copyright Elsevier 2003* [62]

4 Tribological Properties in Carbon Nanotube Polymer Composites

It is unquestionable that if carbon nanotube polymer nanocomposites (CNPN) have been proposed as possible multifunctional materials [13,43], different properties ought to be evaluated. One of the less analyzed properties in CNPN is tribological properties. Nevertheless, even thought few results have been reported in this field, CNPN present diverse and interesting behavior, depending on the different parameters before mentioned, such as: graphitization degree, nanotube structure (zig-zag, armchair, helical) and nanotube kind (multiwalled, single walled or double walled amount others), all directly depending on the nanotubes production method. In addition nanocomposite processing approach, polymer type, modification and additives are extra variables that can improve or diversify carbon nanotube polymer nanocomposite properties. Next we present novel tribological properties presented in CNPN, where, again carbon nanotube loads play an important role, improving and diversifying hardness and wear resistance with respect to the polymer matrix used.

An opposite effect of those discussed above has been reported in CNs nanocomposites, when the hardness has been measured; finding that the increment in the quantity of nanotubes incorporated in the polymer has a favorable effect. The addition of 2 wt % of CNs, improve the hardness of an epoxy matrix in ≈ 20%, however when a less quantity than 1.5 wt% were employed, the hardness lies to diminish in comparison with the hardness of pure epoxy. This effect has been signed on that is caused by a weak interface between CNs and polymer. On the other hand, the favorable effect in the hardness reached whit 2 wt% of CNs, is attributed to the mesh like structures formed when the concentration of CNs is increased. Authors of this research mentioned that the formation of this kind of structures is caused by the high aspect ratio of the nanotubes which produces a tangle arrangement that improve the resistance in the material to be scratched. In addition, in this report different temperatures were used in order to treat CNs nanocomposites and probe their failure in mechanic properties after flexural strength test in diverse conditions, inasmuch as, CNs have been proposed for develop new composites to the space industry and in this context cryogenic states are needed to simulate this kind of environment. Results sign on that the mesh-like structures before mentioned, not improve the flexure strength in the different conditions probed. Authors conclude that in this case, the cause of the reduction could be due to structural non-homogeneity or the presence of a weak-bonding interface [44]. Figure 9. show carbon nanotube mesh like structure produced in this kind of nanocomposites.

In spite of agglomerates and cluster formed in carbon nanotubes polymer nanocomposites lie to diminish some mechanical properties like elastic modulus or rupture strength, recently have been found that this phenomena could produce interesting applications, inasmuch as some anisotropic properties have been found in CNPN produced by solution and casting[32,45]. Other mechanical properties inherent to these nanocomposites were discussed before (see Figure 4 and 5). Nevertheless, these CNPN show other interesting behavior related with their tribological properties. Nanocomposites were analyzed with a micro scratch tester CSEM in progressive mode and Rockwell type, from 0.3 to 20 N of load. The speed was 5.2 mm/min in 4 mm of length. Composites were analyzed by both sides (bottom and top) the bottom side is opaque and the top side is shiny. Figure 10. show a schematic representation of these materials.

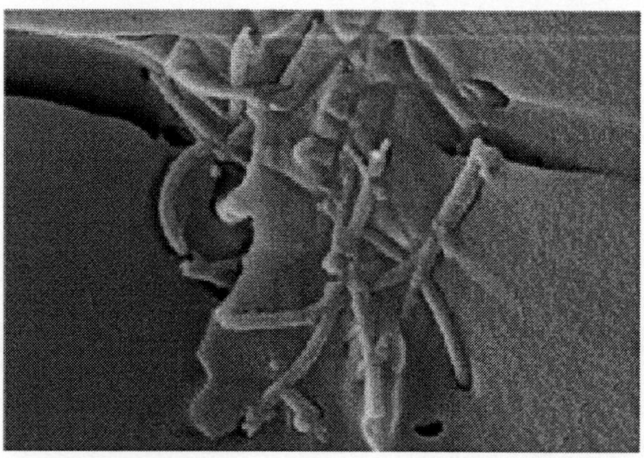

Figure 9. Fracture surfaces parallel to the nanotube mesh in carbon nanotube composite. *Copyright Elsevier 2002*[44].

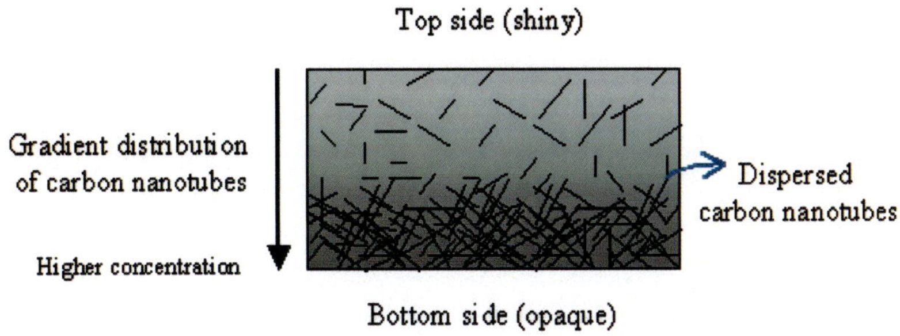

Figure 10. Schematic representation of carbon nanotube polymer nanocomposite with a gradient of distribution.

The analysis has shown that carbon nanotube present different concentration in the polymer. Thus, in spite of these CNPN are very thin, (thickness is less than 0.5 mm) composites have different concentration in top and bottom. This is reflected in interesting anisotropic tribological properties of these materials. Nanocomposites evaluated for wear tests present different behavior depending on which side was evaluated. CNPN and Polymer sample was evaluated in both sides. Figure 11 and 12. shows scratch curves for carbon nanotube composites in opaque and shiny sides respectively. Opaque side is noted with "a" letter and shiny side is distinguished with "b". Number next to letters indicates the carbon nanotube concentration in weight percent. Thus, CNT 1-a, represent CNPN with 1 wt % of carbon nanotubes evaluated in opaque side.

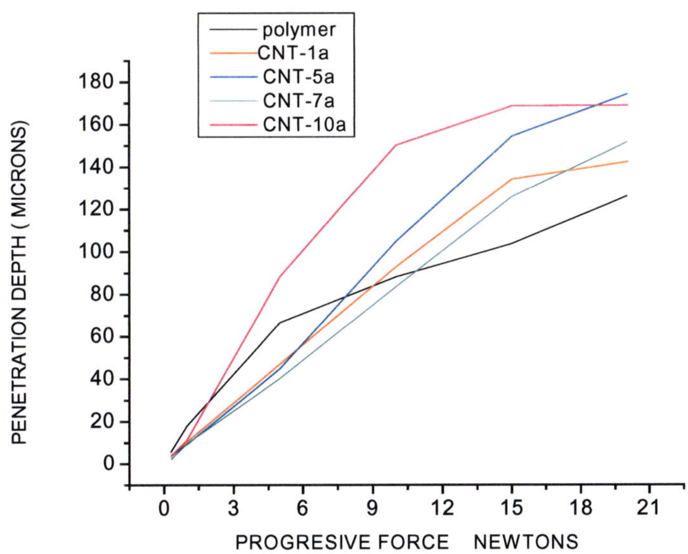

Figure 11. Scratch curves of carbon nanotube polymer nanocomposites bottom side (opaque).

Scratch curves show different behavior depending on the evaluated side. In the figure 11 (opaque side) we can observe penetration depth for polymer sample and CNPN. It's evident that CNPN have different wear resistance depending on the load. At very low load (less than 1.8 N) all CNPN samples present better wear resistance than the polymer matrix. However the sample with 10 wt% diminishes the resistance to be worn drastically, up to 2 N, presenting high penetration depth. The other CNPN samples present better wear resistance than the polymer samples, inasmuch as polymer matrix reach around 67 microns in penetration depth at 5 N of load while CNPN have lower penetration depths at the same load. Composite with 1 wt % of CNs has 45 microns and the composites with 5 and 7 wt % of CNs present 48 and 41 microns of penetration depth at 5 N of load. However, composites present an interesting threshold at high loads where polymer matrix exhibits the same penetration depth at similar loads. CNT-1a presents this point at 9 N with 85 Microns of penetration; CNT-5a samples have this situation around 8.3 N with 84 microns and CNT-7a samples at 11.7 with 93 microns in depth.

In the figure 12 (shiny side) we can observe that penetration depth is around 67 microns at a load of 5 N, for the polymer sample, however at same load CNPN with 1, 5 and 7 wt % show better resistance to be worn. Sample containing 1 wt % of carbon nanotubes shows a penetration depth around 49 microns at 5N of load. At the same load CNT-5b and CNT-7b have depth penetrations around 45 and 43 microns respectively. Nevertheless composites present the threshold at high loads where polymer matrix exhibits the same penetration depth at similar loads. The points in which polymer sample have the same behavior that the CNPN are 10.5 N with 89 microns of depth for CNT-1b, 10.1 N with 88.3 microns for CNT-5b and 10.8 N and 90 microns for the sample CNT-7b. CNT-10b samples present lower wear resistance than the polymer sample at any load.

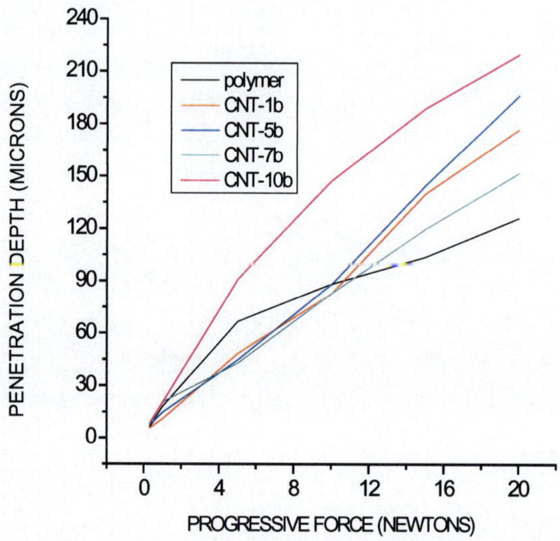

Figure 12. Scratch curves of carbon nanotube polymer nanocomposites top side (shiny).

In both evaluated sides, once that threshold load is reached CNPN with 1, 5 and 7 wt % present lower resistance to be worn than the polymer matrix. The samples present interesting anisotropic wear properties, at low and high load. However near to the threshold zone at medium load composites show similar behavior [46, 47].

These anisotropic properties are caused by the presence of a gradient distribution shown in the figure 10. Thus, although less carbon nanotubes are found in the top side (shiny); these have better distribution in this part of polymer matrix and therefore load transfer is achieve more successfully. Opposite effect was found in bottom side (opaque) where high concentration of nanotubes produces agglomerated which cause holes and poor distribution in different composite regions. Figure 7. shown a Scanning Electron Microscopy (SEM) image in which we can observe these kinds of bundles. In top side also have been found these bundles but in less quantity.

Nevertheless, in spite of these bundles have produced in some cases diminish in mechanical properties of CNPN, mainly in relatively high concentration. The control of the quantity and form of them could be produce interesting anisotropic behavior for composites and other applications. In addition have been signed on recently that these bundles contribute to produce intertube sliding when are incorporated in polymer matrix, and produce interfacial slippage between polymer and nanotubes bundles. This later, producing interesting damping property in carbon nanotube composites[48] and could diversify even more the carbon nanotube nanocomposites properties allowed more possibilities for multifunctional properties and applications.

5 Electrical Conductive Properties in Carbon Nanotube Polymer Nanocomposites

One of the most outstanding properties of carbon nanotubes is their electrical behavior, but once again if they are included in a macroscopic material we can take advantage in different manner their important characteristics for other interesting applications. These can be since electronics devices to automotive and aerospace purposes. Thus, as they are incorporated to different polymers in order to make use of their mechanical properties, research to improve the electrical properties of polymers using carbon nanotubes has been realized also. Different kind of polymers has been used with this aim, from thermoplastics to thermosetting; in almost all researches the reported results have considered the percolation threshold. This parameter is given by the filler concentration at which a pathway or network of conductive filler is formed through the polymeric matrix, and that is clearly observed in changes of conductivity over important quantities.

In composite field epoxy is one of the most used matrix and nanocomposites are not an exception. This polymer is employed in many studies of carbon nanotubes nanocomposites, and besides the mechanical properties developed by the material, they have shown excellent electrical properties with small amounts of carbon nanotubes. Allaoui et al. report a percolation threshold between 0.5 weight % and 1 weight % of carbon nanotubes used as they were obtained from Chemical Vapor Deposition process (CVD) [49]. With similar kind of MWNT Sandler et al. have diminished the percolation threshold, first they found a range of percolation between 0.0225 weight % and 0.04 weight % [50], but later they reduce the percolation threshold until 0.0025 weight %, and found that it is possible to form a carbon nanotubes network since 0.001 weight % [51]. On the other hand Barrau et al. worked also with epoxy but with single walled and double walled nanotubes obtained also from CVD. They report that between 0.2 and 0.4 weight % the composites increase their conductivity by 10 orders of magnitude, the percolation threshold was observed at 0.3 weight %. These authors affirm that the conductivity of the composite is controlled by a percolating path via connecting the carbon nanotubes [52]. In agreement with Sandler they also emphasize the formation of a network of nanotubes within the polymeric matrix.

In other study, Kim et al., using also an epoxy matrix and MWNT, found 5 different percolation thresholds: 0.017, 0.024, 0.028, 0.077 and 0.021 all volume percentages, which depend on the type of oxidative process used in the MWNT purification [53]. They establish that their variety in results are precisely due to the kind of treatment used in the purification of the nanotubes, since the damage to the MWNT depends on the oxidative route, which involves the type of solution (acid or alkaline), its concentration and the temperature and time of treatment. This damage has considerable influence on the electrical properties of the nanocomposites, consequently the electrical conductivity could be lowered at a low content of MWNT and the percolation threshold could be raised.

In the case of thermoplastics, the potential uses of the nanocomposites have increased the number of research made in this sense. For instance, with high density polyethylene (HDPE) He et al [54], found a percolation concentration of 2 weight % whereas for poly methyl methacrylate (PMMA) different researches have found that the threshold goes from 0.003 mass fraction with CVD MWNT [55] to 0.33 mass fraction with arc discharge SWNT [56] . Other thermoplastic polymers nanocomposites are described in table 1.

Not only traditional polymers have been used as matrix for conductive nanocomposites, but also some conjugated polymers improve their conductivity using carbon nanotubes in order to be applied in different electronic devices. Deng et al [66] found that 10 % in carbon nanotubes content increases the conductivity of the composite in more than 25 times than that of the parent polyaniline (PANI).

On the other hand Blanchet et al [67] used also PANI with SWNT synthesized by High Pressure and by Laser ablation. This group obtains a 0.3 % in SWNT content as the conductive threshold. Their results show that conductivity in nanocomposites is highly dependent on the dispersion method but independent of the nanotube fabrication process. Recently Wu et al [68] utilizing the same polymer matrix but with CVD MWNT found that with only 0.1 weight % the conductivity increases dramatically. Other conjugated polymer that was probed as matrix is poly(*m*-phenylenevinylene-*co*-2,5-dioctyloxy-*p*-phenylenevinylene) (PmPv), with it and MWNT from arc discharge Kilbride et al [69] achieved a very low conductive threshold: 0.055 mass %. The table 1 show percolation threshold for other conjugated polymers.

An extensive variety of polymers used as matrix in nanocomposites gives as result many different conclusions about the conductive behavior, even when is the same matrix the processing method to obtain nanocomposites can affect their performance. Other factors that influence conductivity are: the kind of nanotubes (MWNT or SWNT), their production method, dispersion technique and also purification or chemical modification can cause different performances, as were demonstrated above with epoxy nanocomposites.

The low percolation thresholds shown by almost all the different polymer carbon nanotube composites suggest that the type of polymer matrix is not really a determinant factor in the electric response. However the conductive behavior in composites has not been explained clearly and some authors don't agree completely one each other. For instance whereas Barrau et al [52] establish that conduction is mostly achieved by nanotube bundle contacts and so composites behaves as a conductor since there are not nanotubes or bundles separated by polymer barrier, Kilbride et al [69] found precisely the opposite, they assume that nanotubes in composites are probably coated by polymer which acts as a barrier, so the electrical conductivity is limited by tunneling conductive regions.

Table 1. Percolation thresholds for nanocomposites synthesized with carbon nanotubes and different polymer matrix.

Polymer	Percolation threshold	Carbon nanotube type	Reference
Epoxy resin	0.5 – 1.0 wt. %	CVD MWNT	49
Epoxy resin	0.0225 – 0.04 wt. %	CVD MWNT	50
Epoxy resin	0.0025 wt. %	CVD MWNT	51
Epoxy resin	0.3 wt. %	CVD SWNT – DWNT	52
Epoxy resin	0.017 – 0.077 vol. %	CVD MWNT	53
Epoxy resin	3.0 wt. %	SWNT	57
Epoxy resin	0.074 wt. %	EAD SWNT	58
High density polyethylene	2.0 wt. %	CVD MWNT	54
Poly(methyl methacrylate)	0.003 mass fraction	CVD MWNT	55
Poly(methyl methacrylate)	0.33 mass %	EAD SWNT	56
Polycarbonate	1.0 – 1.5 wt. %	MWNT	59
Polycarbonate	1.0 – 2.0 wt. %	MWNT	60
Polyurethane	0.005 vol. fraction	MWNT	61
Polyimide	0.1 vol. %	Laser SWNT	62
Polystyrene	1.0 wt. %	CVD MWNT	63
Poly(*p*-phenylenevinylene-*co*-2,5-dioctyloxy-p-Phenylenevinylene)	8.4 mass fraction	EAD MWNT	64
Poly(styrene-co-butyl acrylate)	≈ 2.5 wt. %	CVD MWNT	65
Polyaniline	10 %	No specified	66
Polyaniline	0.3 wt. %	HP SWNT Laser SWNT	67
Polyaniline	0.1 wt %	CVD MWNT	68
Ethyl cellulose	3.0 wt. %	HP SWNT Laser SWNT	67
Poly(*m*-phenylenevinylene-*co*-2,5-dioctyloxy-p-Phenylenevinylene)	0.055 mass %	EAD MWNT	69
Polyvinyl alcohol	0.055 mass %	EAD MWNT	69
Poly(phenylene ethynylene)/ polystyrene	0.045 wt. %	HP SWNT	70
Poly(phenylene ethynylene)/ polycarbonate	0.11 wt. %	HP SWNT	70
Poly(3-octylthiophene)	11.0 wt. %	EAD SWNT	71

EAD= Electric Arc Discharge
CVD= Chemical Vapor Deposition
HP= High Pressure of Carbon Monoxide (HiPCO)

Another important feature in the nanocomposite conductive performance is the alignment of carbon nanotubes inside the matrix, which depending on the area in where they are set can develop an anisotropic behavior in conductivity. Recently Martin et al [72] used an electric field to induce alignment in MWNT in epoxy and even though the network efficiency was not improved, they said that the observed orientation of the field-induced nanotube networks shows promising optical transparency and should yield anisotropic electrical properties. However Watts et al [73] were one of the first groups in report anisotropic electrical conductivity. In that article they synthesized polystyrene with MWNT and boron doped MWNT and defined as top the face of the composite exposed to the vacuum and as bottom the face in contact with Petri dish. The bottom face of the film had a higher conductivity compared with the top. This difference, they assumed, is due to the presence of a gradient distribution of the carbon nanotubes through the polymer matrix possibly due to particle settlement during composite formation.

Recently have been found other interesting results where a similar phenomenon is presented; however in this research were not used doped nanotubes. The CNPN are those discussed in previously sections. These composites were prepared by solution and casting[32, 45] and also shown interesting tribological anisotropic behavior discussed in the last section 46, 47. The effect that produce distribution gradient in these materials was shown in the figure 10. This phenomenon cause different conductive properties in carbon nanotube composites. As we mentioned before the bottom side have more concentration of nanotubes which produce conductive properties. However the concentration of nanotubes presented in top side does not produce conductive behavior. The samples with 7 and 10 wt % of nanotubes present low conductivity in top side. Nevertheless, bottom side presents more electrical conductive properties than top side in all concentrations. Thus, these materials present interesting anisotropic electrical conductive properties inherent to different grade of dispersion in polymer matrix and holes produced in the interface of composites close to the carbon nanotubes bundles (figure7). These zones create a barrier effect between top and bottom sides producing different conductive behavior. Figure 13 illustrates conductive behavior of bottom side; extra composite with 3 wt % of nanotubes was included for this analysis. In addition the figure presents a comparison with other polymer that shows similar percolation mechanism in electrical conductive properties of CNPN[74-75].

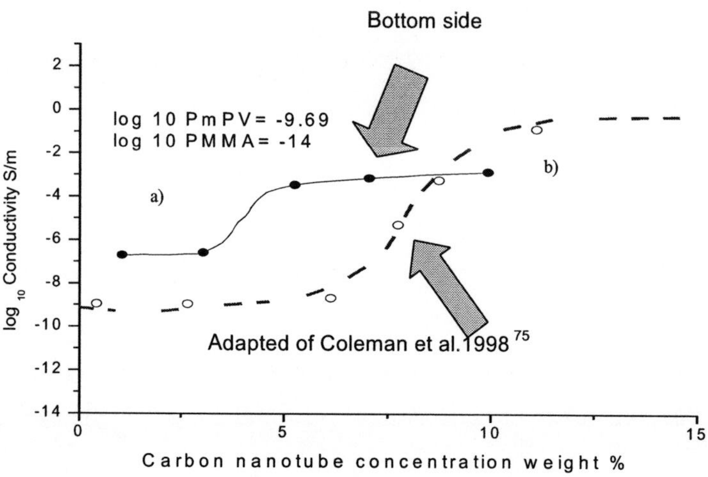

Figure 13. Conductivity behavior in two carbon nanotube polymer nanocomposites. a) poly (methyl-ethyl methacrylate) reinforced with MWNTs. b) PmPV *Poly(p-phenylenevinylene-co-2,5-dioctyloxy-m-phenylenevinylene)* reinforced with MWNTs.[75] (log 10 Conductivity of PmPV and PMMA are presented as reference)

Research presented in this section show that carbon nanotube polymer nanocomposites show interesting electrical properties, where we can take advantage of not only low percolation threshold but in addition novel anisotropic conductive properties could be useful for different applications. Investigation in this field is growing and diverse behaviors related to electrical conductivity in CNPN are found frequently allowing diversifies CNPN uses.

6 Chemical Functionalization of Carbon Nanotubes

The results analyzed in the firsts section show that CNs obtained from production methods that produce impurities together with CNs, develop sometimes unpredictable properties, inasmuch as even though the composites in some cases show an important increase in the mechanical properties, the effect is reduced when more than 5% of CNs are incorporated in polymer. In addition other papers report not so good interface due to a poor wettability[76] which could be influenced by the impurities in some cases. So although have been found that could be possible incorporate low load of these kinds of nanotubes without reduce drastically mechanical properties and take advantage of other properties like damping; there exist other possible applications where carbon nanotube with other features could be needed. Thus, different methods have been proposed with the target to eliminate impurities in CNs. The most important impact has been produced by oxidation methods which in addition to reduce impurities cause chemical modifications on CNs. This has been proposed as a useful tool in the area of composites and in other applications of these materials. This section describes some important studies in the field of chemical modification, which could be useful to improve the links between CNs and polymer matrix with the aim to diversify and enhance the performance of polymer nanocomposites.

The oxidation has opened a new gate in this field, inasmuch as produces carboxyl and carboxylate groups in the tips and in the surfaces of CNs. These moieties have been used in order to attach different chains in CNs that allow improving different possible uses of these materials[77]. Oxidation approaches were developed with different aims few years later CNs were discovered. The approaches took as base the different resistance to be oxidized between CNs and other nanoparticles formed during the production. Thus, it was observed that the nanotubes when are oxidized are consumed from the tips inward at the same time[78] and the surface wall is modified if the oxidation is severe.

Oxidation in the surface of these materials have been demonstrated to produce useful moieties in order to bond new reactive chains that improve solubility, processability and compatibility with other materials and therefore improve the interface properties of CNs with other substances, allowing to take advantage in better manner of CNs properties. A amply description of oxidation and chemical modification in carbon nanotubes was describe in reference [79].

The appearance of defects caused by oxidation has been studied, finding that nanotubes tolerate a limited quantity of them, before a macroscopic sample loses its special mechanical and electronic properties [80-82]. Organic carboxyl groups formed in nanotube surface are localized in the defects on functionalized single-walled and multiwalled nanotubes (f-SWNTs, f-MWNTs respectively); suitable reactive organic groups for other chemical chains lie to react in this zone. A representative scheme functionalized CNs with these organic moieties is shown in the Figure 14 [13].

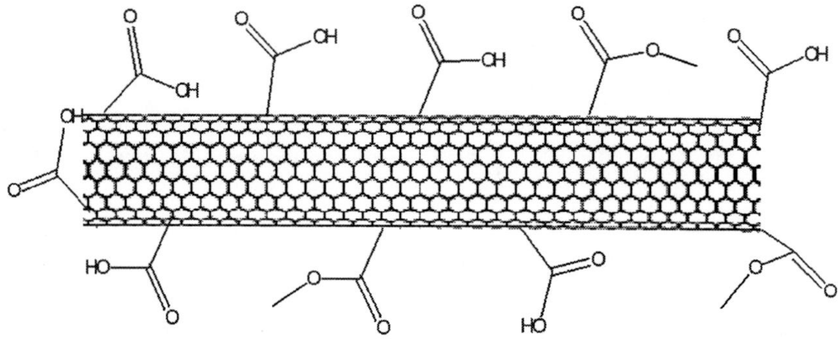

Figure 14. A schematic representation of oxidized and functionalized carbon nanotube. *Copyright Brill Academic Publishers 2005* [13]

Next are presented and discussed different researches of chemical functionalization after a previous oxidation. In these papers the COOH groups generated in the oxidation process are used in order to attach different molecules which have been useful to improve surface compatibility of CNs with other materials. The relevance of these groups to join CNs with organic polymer matrices is also mentioned.

One of the routes that has taken part as complementary procedure to insert long organic chain is realized with $SOCl_2$ (thionylchloride) in dimethyl formamide, thus long chain alkylamine (Octadecylamine) has been coupled to activated f-SWNTs.[83]. This method is very useful to insert different chains and therefore is useful to improve compatibility with polymer matrices. The approach also has been used to enhance the solubility properties of SWNTs with the mentioned chains and other moieties as alkyl-aryl amine and 4-dodecyl-aniline [84].

Other researches have been developed with a wide variety of aliphatic amines coupled to carboxyl groups in SWNT surface [85], showing that amines in general can be attached to CNs successfully. The insertion of these groups in CNs surface could be useful to improve compatibility between these nanomaterials and aminopolymer matrices.

Dodecylamine groups have been bonded to CNs through carboxyl moieties via $SOCl_2$ reaction. The functionalized carbon nanotubes have shown an improvement in the solubility with polymer matrix and organic solvents and have been used to develop carbon nanotube-oxotitanium phthalocyanine composite. The material have presented improve in the photosensitivity with the content of functionalized CNs and an important increase of five fold higher than the only oxotitanium phthlocyanine [86].

The carboxylic terminated organic groups on SWNTs have been used by Quin et al.[87] in order to attach n-butylmethacrylate via atom transfer radical polymerization. In this reaction the polymerization is controlled and might be useful in order to graft copolymers blocks and other polymers, making better the compatibility between nanotubes and this kind of synthetic macromolecules.

In the context of improve CNs compatibility with polymers different researches have been developed recently and other studies are in progress. A recent research promotes the insertion of many functional groups on CNs surface. Using the same reaction, different organosilanes were added to MWNT via carboxyl groups on modified CNs. The method allows diversifying CNs properties depending on the terminal chain of the coupled organosilane. The oxidation and silanization realized in this research yield functionalized moieties, which contain terminated organosilanes groups denominated as R groups that are bonded to the nanotubes by silanol groups. This has allowed modifying organic-terminated CNs with many different chains chemically-bonded to MWNTs, with silanization reaction, inasmuch as the R group (methacryl, glycidoxy, etc) in the organofunctional silanes has been changed in current research depending on the specific polymeric matrix employed for a particular use. The behavior of these MWNTs changes when the R terminal changes, so it has been found that 3-Methacryl-trimethoxysilane (3-MAT) produces soluble MWNTs in organic solvent such as acetone or ethanol.[45, 88]

One of the most novel and promising techniques to improve the compatibility is polymer grafting on the surface of carbon nanotube walls. Many of the techniques used with this aim are taken from polymer grafting on natural fibers, however almost all need modifications accordingly with the nature of CNs. At the same time some of the functional moieties attached as were described before are very useful in order to prepare surfaces for grafting. However there is a huge diversity of chemical routes for grafting, mainly due to it is possible use the classic organic type reactions, in combination or with modifications, and of course design new arrangements of synthesis via.

A classic cycloaddition reaction on SWNT treated with HNO_3, preceded by free radical and substitution reactions that prepare a nitrogen end group polystyrene (-CH(Ph)-N3), is used to made a "grafting to" procedure [89]. This allows the growing of polystyrene chains on the grafted side wall nanotubes. A schematic representation of this procedure is shown in figure 15.

On the other hand atom transfer radical polymerization (ATRP) could be useful to graft different polymers from modified nanotubes, as was demonstrated by Qin et al [89, 90]. They made a "grafting from" method to insert polystyrene and n-Butyl Methacrylate on SWNT modified with bromopropionate groups. A brief summary of the conducted reactions is presented in figure 16.

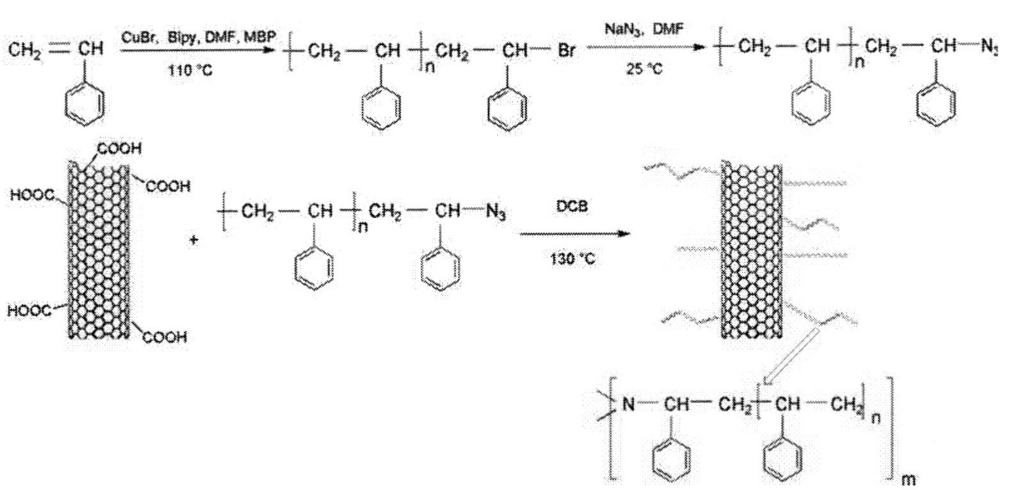

Figure. 15. Schematic diagram of "grafting to" reaction to attach Polystyrene on CNs, where the necessary substances are copper bromide (CuBr), 2,2'-bipyridine (Bipy), N,N-Dimethylformamide (DMF), Methyl-2-bromopropionate (MBP), sodium azide (NaN_3) and 1,2- dichlorobenzene (DCB).

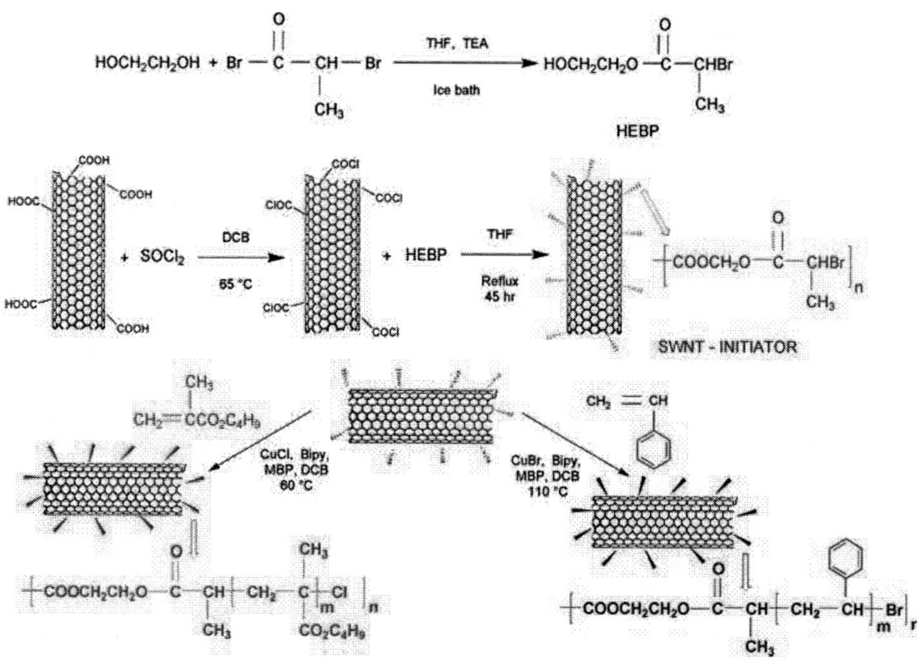

Figure. 16. Schematic representation of “grafting from” reaction to attach Polystyrene and n-Butyl Methacrylate on CNs. Some of the necessary substances were listed before, but also are needed: tetrahydrofuran (THF), triethylamine (TEA), thionyl chloride ($SOCl_2$), copper chloride (CuCl) and 2-hydroxyethyl-2-bromopropionate (HEBP).

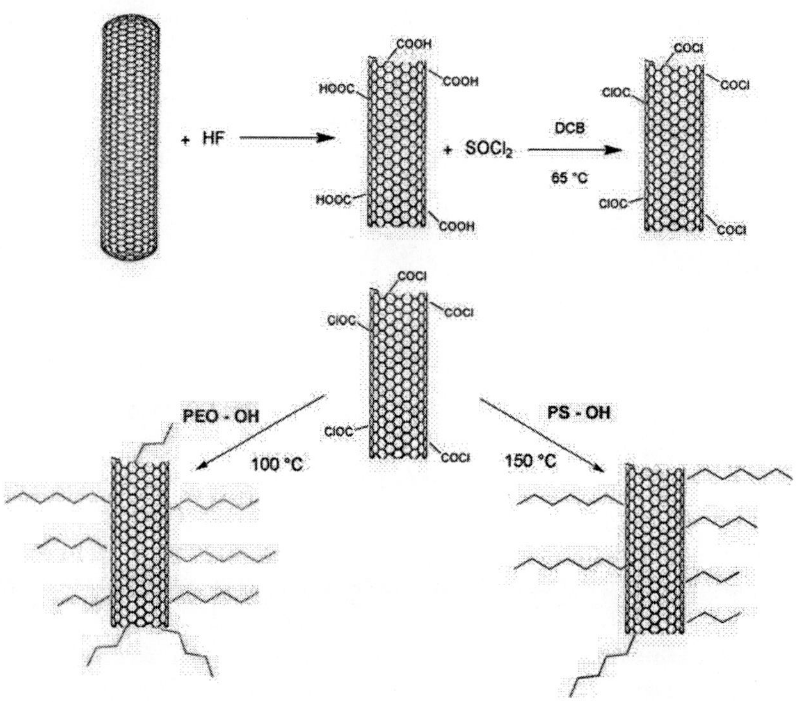

Figure 17. Schematic reaction of grafting Polystyrene and Poly (ethylene oxide) using hydroxyl terminated polymers to attach the CNs.

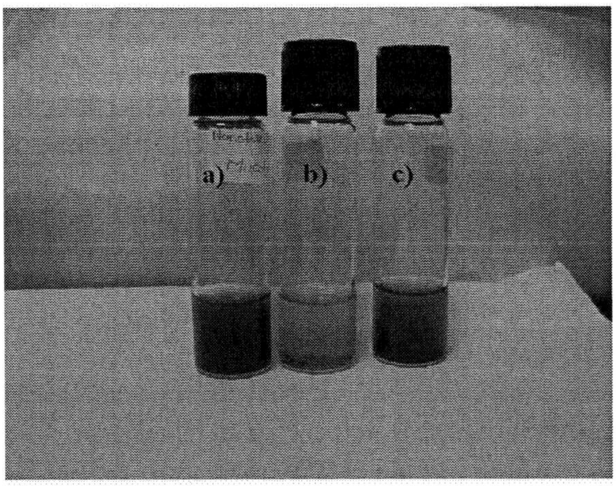

Figure 18. Carbon nanotubes dissolved in acetone, a) Grafted PMMA Multiwalled carbon nanotubes (g-MWNTs) , b) oxidized functionalized single walled carbon nanotubes (f-SWNTs), c) sample 2 of grafted pmma multiwalled carbon nanotubes (g-MWNTs). Images took after 24 hrs without agitation.

A different grafting reaction system to attach polystyrene and poly(ethylene oxide) was published recently by Baskaran et al [91]. It is based in the covalent functionalization of MWNTs through the esterification reaction of terminal hydroxyl polymers. This system uses the hydroxyl group to be joined with MWNTs – COCl produced by HF purification followed by a reaction with $SOCl_2$. Figure 19 shows a schematic diagram of this reaction system.

The most new techniques in organic chemistry can be applied also in grafting CNs, for instance Radical Polymerization with Reversible fragmentation chain transfer (RAFT polymerization) which is one of the most modern procedure for preparation of polymers is used to bond polystyrene to MWNTs [92]. The actually classic chemical functionalization with HNO_3 is used to yield MWNT – COOH, later MWNT - COCl were produced with the addition of $SOCl_2$. The RAFT agent is produced by the reaction between MWNT – COCl and 2-hydroxyl-ethyl-2'-bromoisobutyrate. Finally the MWNTs functionalized as RAFT agent begin the addition reaction with 2-2 azobisisobutyronitrile (AIBN) and styrene monomer, the products of this reaction are polystyrene grafted MWNTs.

An interesting research of grafting but without damage the outer walls or end caps of CNs was developed by Viswanathan, et al [93]. Their novel method is realized by in situ polymerization through charge transference from Butyllithium to CNs, producing a carbanion, that later attacks the double bond of styrene monomer. Successive addition completes the grown of polystyrene attached to CNs.

In general chemical functionalization has been used successfully to improve compatibility with polymer matrix, and grafting is other important route exploited recently. The grafted chains inserted in nanotubes have demonstrated that it is possible to change the surface properties in nanotubes. This allows diversifying properties in similar manner that those occurred with other chemical modifications mentioned previously. In fact grafting reactions have produced the insertion of other polymer in carbon nanotube surface. Other grafted monomer is methylmethacrylate. Solubility properties of grafted nanotubes with this polymer are presented in figure 18. The reaction used in this research is a modification of polymer grafting on a natural fiber [94]. More details of these grafted nanotubes will be presented extensively in reference [95].

Thus, chemical functionalization is a very important route in order to be compatibly nanotubes and polymer matrix. This has allowed producing new synergic composites. Next are presented some researches in CNPN using functionalized nanotubes.

7 Functionalized Carbon Nanotube-Polymer Nanocomposites

Chemical modification is not a new tool to enhance the compatibility between reinforcements and matrix, inasmuch as chemical attack and graft have been used in natural fibers, synthetic fibers and particles [94, 96, 97]. However the carbon nanotube nanocomposite field is very recently and chemical functionalization offers an important tool in order to improve interface and reach the amazing properties that carbon nanotubes can provide. Moreover although different researches have taken part in this field, it is needed to understand the possible interactions in order to control and to improve the interface between carbon nanotubes and polymer matrix. The recent results in chemical modification of carbon nanotubes to incorporate these materials to polymer are promising and could be the way to form strong and functional new advanced materials. Some results obtained in the area of functionalized carbon nanotube nanocomposites are described in this section.

In previous section we have reviewed that the carboxyl groups found in the tip and surface of CNs are useful for the interaction between CNs and other compounds. Thus, the use of these moieties to improve the "link" in CN-polymer composites has been proposed either inserting other chemical groups or using the carboxyl groups produced in the oxidation[88]. Jia et al. [37] suggested that the initiator opens the π bonds found in CNs and, in this way, CNs take part in the polymerization. However have been signed on that although opening π bonds in CNs to join them with other chemical groups has been proposed, oxidation provides more possibilities to bond the nanotubes with the matrix, due to reactive chemical groups such as COOH, COO- and C=O which are found on the tip and on the wall surface. In another report Geng et al.[98] use fluorinated single-walled nanotubes (fl-SWNTs) to enhance the uniformity and the nanotube dispersion using poly(ethylene oxide) (PEO) as matrix, and dissolving the nanotubes in methanol.

Other researches [33] have proposes that definitively, the best moment to reach possible interactions between the functional groups (found in tips and walls in CNs) and polymer chains is when the polymer is created by in situ processing, inasmuch as the free radical formed in monomer molecules by the initiator could either interact or react with the CN moieties easier than when the polymer is made and after dissolved or melted to produce the composites. In that research Methyl Methacrylate monomer (MMA), 2-2 azobisisobutyronitrile (AIBN) and oxidized f-MWNT were used, and CN composites were produced *"in situ"* polymerization using AIBN as initiator. The AIBN quantity, reaction time and temperature were controlled to have uniform molecular weights in all samples. Three samples in this research were manufactured with only polymer (poly (methyl methacrilate PMMA)): polymer with 1 wt % of unfunctionalized MWNTs (u-MWNTs), and polymer with 1 wt % of functionalized MWNTs (f-MWNTs), additional composite with polymer and 1.5 wt% of f-MWNTs was also produced to observe the behavior when the addition of f-MWNTs was increased slightly. In the research was proved the interaction of PMMA with f-MWNTs by Infrared and Raman spectroscopy. In addition mechanical result shown an important increase in E' at 40 °C by 66 % and 88 %, with 1 wt% and 1.5 wt% of f-MWNTs respectively, and both samples increase the glass transition temperature Tg by ≅ 40 °C, unlike with unfunctionalized nanotubes where E' at 40°C increase by 50 % and the Tg by only 6 °C. Another important fact found in this paper in that f-MWNTs composites increase more than 11-fold, at relative high temperature. A comparison of different properties of this and other f-CNs composites is presented in the reference [79].

Other recent papers have shown that carboxyl-terminated carbon nanotubes treated with amines and incorporated to epoxy matrix have better interaction than those that were not treated chemically, showing an important improvement in the interface links between these two materials. In this study TEM pull out test were developed showing a better interaction due to chemical groups on CNs surface [99]. Other authors have develop in situ functionalized MWNTs in phenoxy composites by melt mixing with 1-(aminopropyl)imidazole [100]. In this research the composites with more than 4.8 wt % show higher storage modulus than the polymer matrix, however the behavior in high temperature is not common in these materials, inasmuch as f-MWNTs lie to increase the modulus when normally the polymer matrix lie to flow. In this research the modulus at high temperature is diminished with the incorporation of in situ functionalized nanotubes.

In fact functionalized carbon nanotube polymer nanocomposites are developed by different approaches and with different chemical moieties. Results have shown important increments in storage modulus and glass transition temperature between other important parameters for example hardness. In recently research Owens J.[101] used fluorinated and non fluorinated SWNTs included in Polyacrylonitrile; Hardness results of fluorinated nanotube composite present more hardness than non fluorinated nanotube composite. For instance a sample which contain 12 wt % of fluorinated nanotubes have a hardness around 100 according with ASTM D2240 while the composite with the same quantity of non-fluorinated nanotubes present a hardness less than 80. Similar effects were obtained with less concentration of functionalized and non functionalized nanotubes in composites.

Liu L. and D. Wagner [102] recently produce two CNPN developed with two types of epoxy resins using f-MWNT and MWNTs. f-MWNTs consisted of carboxyl terminated nanotubes. Viscosity was a determined point in the synthesis of these composites, inasmuch as better dispersion was reached in rubbery epoxy resin than in glassy epoxy. However in both cases functionalized carbon nanotube nanocomposites present better tensile modulus than those produced with MWNTs and the only resins, using 1 wt % of nanotubes. In addition very good impact properties were reached when functionalized nanotubes were incorporated to epoxy resin.

Other researches have used carboxylated terminated nanotubes in order to reinforce polyamide 6 [103] and MWNTs modified with ferritin protein molecules to reinforce Poly (vinyl-alcohol) PVA [104]. In both researches were used also un-functionalized nanotubes. The storage modulus and glass transition temperature were incremented more notable when functionalized nanotubes were incorporated in polymer matrix. Other functionalized terminated SOH nanotubes have been used to reinforce PVA polymer matrix improving mechanical properties [105] and carboxyl terminated nanotubes in polyurethane have produce interesting electroactive shape recovery promising in the field of actuator materials [106].

Recently non-covalent functionalization in carbon nanotubes was used for produce a new ultra strong composite employed to create the first bicycle where the frame have as reinforcement carbon nanotubes. BMC the company that produced the bicycle employed polymer composite material developed between Zyvex which provide functionalized nanotubes and Easton [107]. This represents the first CNPN that is developed for special application, however in concordance with all mentioned in these section. Chemical functionalized nanotube nanocomposites have open different important field to diversify and improve carbon nanotube polymer composite properties and applications.

8 Conclusions

Carbon nanotube polymer nanocomposites promise to be an important field that could develop a new age of ultra strong, light, conductive and versatile materials. These features in one material may origin new important multifunctional nanocomposites. However in spite of the relevant and novel results of these materials, research in this field is beginning and even is needed to understand many aspects of CNPN, inasmuch as several of the analyzed reports are in agreement that CNs in large amounts lie to form clusters in some areas. This diminishes the interaction and therefore the interface is not the best. Nevertheless, manage of some of these parameters could be diversify nanocomposite properties as we analyzed in this chapter.

The incorporation of few quantities of nanotubes in polymer matrices improve in significantly manner the properties, reaching important values never seen with other materials in electrical, thermal, optical and mechanical properties.

In addition chemical functionalization is nowadays one of the most useful approaches to be compatibles carbon nanotubes with polymer matrix and with other materials in nanotechnology field, improving different properties of CNPN. Thus, comprehension of the ideal conditions in functionalization of CNs in order to improve interactions and understand the interface in CNs nanocomposites could produce advanced materials with unexpected properties. As well, diversification produced by different chemical functionalization might to be real the promising age of multifunctional materials, inasmuch as the synergetic effect that produce modified nanotubes with organic polymer matrix generate properties that are not available in any other material

Acknowledgments

We thank to Mr. D. Rangel and Mr. E. Mendez for their technical assistance in different measures of some results presented in this chapter.

The kind authorization for using some figures from Elsevier Science and American Institute of Physics, and a brief text from Brill Academic Publishers is gratefully acknowledged.

References

[1] Iijima, S. *Nature*, **354**, 56, (1991).
[2] M.M.Treacy, T.W. Ebbesen and J.M. Gibson, *Nature*, **381**, 678 (1996).
[3] O.Lourie and H.D.Wagner, *J. Mater. Res.* **13**, 2418 (1998).
[4] O.Lourie, D.M.Cox and H.D.Wagner, *Phys. Rev. Lett.*, **81** 1638 (1998).

[5] M-F. Yu, O. Lourie, M.J. Dyer , K. Moloni , T.F. Kelly and R.S. Ruoff, *Science* **287**, 637 (2000).
[6] K. I. Tserpes and P. Papanikos, *Compos. Part B: Eng.* **36**, 468 (2005).
[7] T.W. Ebbesen, H.J. Lezec, H. Hiura, J.W. Bennett, H.F. Ghaemi and T. Thio. *Nature,* **382**, 54 (1996).
[8] J. Che, T. Cagin and W.A. Gooddard III, *Nanotechnology,* **11**, 65 (2002).
[9] L. An, W. Xu, S. Rajagopalan, C. Wang, H. Wang, Y. Fan, L. Zhang, D. Jiang, J. Kapat, L. Chow, B. Guo, J. Liang and R. Vaidyana, *Adv. Mater.* **16**, 2036 (2003).
[10] G-D. Zhan and A. K .M. Mukherjee, *Int. J. Appl. Ceram. Technol.* **1**, 161 (2004).
[11] G-D. Zhan, J. D. Kuntz, H. Wang, C-H. Wang and A. K .M. Mukherjee, *Phil. Mag. Lett.* **84**, 419 (2004).
[12] R. George, K. T. Kashyap, R. Rahul and S. Yamdagni, *Scripta Mater.* **53**, 1159 (2005).
[13] C. Velasco-Santos, A.L. Martinez-Hernandez, and V.M Castano, *Comp. Interfaces* **11**, 567 (2005).
[14] T.W. Ebbesen and P.M. Ajayan, *Nature*, **358**, 220 (1992).
[15] C. Journet, W. Maser, B. Bernier, A. Loiseau and M. de la Chapelle. *Nature,* **388**, 756 (1997).
[16] W.Z. Li, S.S. Xie and G. Wang. *Science,* **274**, 1701 (1996).
[17] Thess, R. Lee, P. Nokolaev, H. Dia, and P. Petit, *Science,* **273**, 483 (1996).
[18] W.K. Hsu, M. Terrones, H. Terrones, N. Grobert, A. I. Kirkland, J. P. Hare, K. Prassides, P. D. Townsend, H. W. Kroto and D. R. M. Walton, *Chem. Phys. Lett.* **284**, 177 (1998).
[19] L. S. Wojciech, J.A. Libera, Y. Gogotsi and Y. Masahiro, *J. Solid. State. Chem.* **160**, 184 (2001).
[20] Velasco-Santos, A.L. Martínez-Hernández, A. Consultchi, R. Rodríguez and V.M. Castaño, *Chem. Phys. Lett,* **373**, 272 (2003).
[21] P.M. Ajayan, O. Stephan, C. Colliex and D. Trauth, *Science*, **265**, 1212 (1994).
[22] L. Jin, C. Bower and O. Zhou *Appl. Phys. Lett.* **73**, 1197 (1998).
[23] R. Haggenmueller, H.H. Gommans, A.G. Rinzler, J.E. Fischer and K.I. Winey *Chem. Phys. Lett.* **330**, 219 (2000).
[24] H. D. Wagner, O. Lourie, Y. Feldman and R. Tenne, *Appl. Phys. Lett.* **72**, 188 (1998).
[25] O. Lourie and H.D. Wagner, *Compos. Sci. Technol.* **59**, 975 (1999).
[26] F.T. Fisher, R.D. Bradshaw and L.C. Brinson, *Appl. Phys. Lett.* **80**, 4647 (2002).
[27] F.T. Fisher Carbon nanotube reinforced polymers. Ph. Thesis. Northwestern University. Evanston Illinois. USA. (2002).
[28] Qian and E.C. Dickey, *J. Microsc.* **204**, 39 (2001).
[29] Stephan, T.P. Nguyen, M. L. De la Chapelle, S. Lefrant, C. Jounet and P. Bernier, *Synth. Met.* **108**, 139 (2000).
[30] M.L. de la Chapelle, C. Sthepan, T.P. Nguyen, S.Lefrant, C. Jounet, P. Bernier, E. Munoz, A. Benito, W. K. Maser, M.T. Martinez, G.F. de la Fuente, T. Guillard, G. Flamant, L. Alvarez and D. Laplaze, *Synth. Met.* **103**, 2510 (1999).
[31] Z. Jin, K.P. Pramoda, G. Xu and S. Hong Goh, *Chem. Phys. Lett.* **337**, 43 (2001).
[32] C. Velasco-Santos, A.L. Martínez-Hernández, F. Fisher, R.S. Rouff and V.M. Castaño, *J. Phys. D. Appl. Phys.* **36**, 1423, (2003).
[33] C. Velasco-Santos, A.L. Martínez-Hernández, F. Fisher, R.S Ruoff and V.M Castaño, *Chem. Mat.* **15**, 4470 (2003)

[34] M.L. Shofner, F.J. Rodriguez-Macias, R. Vaidyanathan and E.V Barrera, *Comp. Part. A*. **34**, 120 (2003).

[35] L.S. Schadler, S.C. Giannaris and P.M. Ajayan, *Appl. Phys. Lett.* **73**, 3842 (1998).

[36] M.S.P Shaffer and A. H. Windle *Adv. Mater.* **11**, 937 (1999).

[37] Z. Jia, Z. Wang, C. Xu, J. Liang, B. Wei, D. Wu and S. Zhu. *Mater. Sc. Eng.* **A271** 395 (1999).

[38] C. A. Cooper, D. Ravich, D. Lips, J. Mayer and H.D. Wagner, *Compos. Sci. Technol.* **62** (2002) 1105.

[39] S.R: Tatro, L.M. Clayton, P.A. O'Rourke Muisener, A.M. Rao and J.P. Harmon. *Polymer,* **45**,1971 (2004).

[40] K.W. Putz, C.A. Mitchell, R. Krishnamoorti and P.F. Green, *J. Polym. Sci. B.* **42**, 2286 (2004).

[41] Dufresne, M. Paillet, J.L. Putaux, R. Canet, F. Carmona, P. Delhaes and S. Cui, *J. Mater. Sc. Lett.* **37**, 3915 (2002)

[42] M. Wong, M. Paramsothy, X.J. Xu, Y. Ren, S. Li and K. Liao, *Polymer,* **44**, 7757 (2003).

[43] Velasco-Santos, A.L. Martínez-Hernández and V.M Castaño, *Technical Proceedings, Nanotech,* **2**, 139 (2005).

[44] K-T. Lau and S-q. Shi, *Carbon,* **40**, 2965 (2002).

[45] Velasco-Santos Ph. D. Thesis, Universidad Autónoma de Querétaro – CFATA UNAM. México (2003).

[46] Velasco-Santos, A.L. Martínez-Hernández, W. Brostow and V.M Castaño, *Presented in CNT Polymer Internacional Conference, Hamburg, Germany* (2005).

[47] C. Velasco-Santos, A.L. Martinez-Hernandez, and V.M Castano In preparation for *Compos. Sci. Technol.* (2005).

[48] J. Suhr, N. Koratkar, P. Keblinski and P. Ajayan, *Nature Mat.* **4**, 134 (2005).

[49] Allaoui, S. Bai, H.M. Cheng and J.B. Bai, *Compos. Sci. Technol.* **62**, 1993 (2002).

[50] J. Sandler, M.S.P. Shaffer, T. Prasse, W. Bauhofer, K. Schulte and A.H. Windle, *Polymer,* **40**, 5967 (1999).

[51] J.K.W. Sandler, J.E. Kirk, I.A. Kinloch, M.S.P. Shaffer and A.H. Windle, *Polymer*, **44**, 5893 (2003).

[52] S. Barrau, P. Demont, A. Peigney, C. Laurent and C. Lacabanne *Macromolecules*, **36**, 5187 (2003).

[53] Y.J. Kim, T. S. Shin, H. D. Choi, J. H. Kwon, Y-C. Chung and H. G. Yoon. *Carbon*, **43**, 23 (2005).

[54] X. J. He, J. H. Du, Z. Ying and H. M. Cheng, *Appl. Phys. Lett.* **86**, 062112 (2005).

[55] H.M. Kim, K. Kim, C.Y. Lee, J. Joo, S.J. Cho, H.S. Ion, D.A. Pejakovic, J.W. Yoo and A.J. Epstein, *Appl. Phys. Lett.* **84**, 589 (2004).

[56] J-M. Benoit, B. Corraze, S. Lefrant, W.J. Blau, P. Bernier and O. Chauvet, *Synth. Met.* **121**, 1215 (2001).

[57] E.S. Choi, J.S. Brooks, D.L. Eaton, M.S. Al-Haik, M.Y. Hussaini, H. Garmestani, D. Li and K. Dahmen, *J. Appl. Phys.* **94**, 6034 (2003).

[58] Kim, J. Lee and I. Yu, *J. Appl. Phys.* **94**, 6724 (2003).

[59] P. Pötschke, A.R. Bhattacharyya and A. Janke, *Carbon,* **42**, 965 (2004).

[60] P. Pötschke, T.D. Fornes and D. R. Paul, *Polymer*, **43**, 3247 (2002).

[61] H. Koerner, W. Liu, M. Alexander, P. Mirau, H. Dowty and R.A. Vaia, *Polymer* **46**, 4405 (2005).
[62] Park, Z. Ounaies, K.A. Watson, R.E. Crooks, J. Smith Jr., S. E. Lowther, J. W. Connell, E.J. Siochi, J.S. Harrison and T. L. St. Clair, *Chem. Phys. Lett.* **364**, 303 (2002).
[63] B. Safadi, R. Andrews and E. A. Grulke, *J. Appl. Polym. Sci.* **84**, 2660 (2002).
[64] H. J. Barraza, F. Pompeo, E. A. O'Rear and D. E. Resasco, *Nano. Lett.* **2**, 797 (2002).
[65] J.N. Coleman, S. Curran, A.B. Dalton, A.P. Davey, B. McCArthy, W. Blau and R.C. Barklie, *Synth. Met.* **102**, 1174 (1999).
[66] J. Deng , X. Ding , W. Zhang , Y. Peng, J. Wang, X. Long , P. Li and A.S.C. Chan, *Eur Polym. J.* **38**, 2497 (2002).
[67] G. B. Blanchet, C. R. Fincher and F. Gao, *Appl. Phys. Lett.* **82**, 1290 (2003).
[68] T-M. Wu, Y-W. Lin and C-S. Liao, *Carbon*, **43** 734 (2005).
[69] B. E. Kilbride, J. N. Coleman, J. Fraysse, P. Fournet, M. Cadek, A. Drury, S. Hutzler, S. Roth and W. J. Blau *J. Appl. Phys.* **92**, 4024 (2002).
[70] R. Ramasubramaniam J. Chen and H. Liu, *Appl. Phys. Lett.* **83**, 2928 (2003).
[71] Kymakis, I. Alexadou and G.A.J. Amaratunga, *Synth. Met.* **127**, 59 (2002).
[72] C.A. Martin, J.K.W. Sandler, A.H. Windle, M.-K. Schwarz, W. Bauhofer, K. Schulte and M.S.P. Shaffer, *Polymer,* **46** 877 (2005).
[73] P.C.P. Watts, W.K. Hsu, G. Z. Chen, D. J. Fray, H. W. Kroto and D.R.M. Walton, *J. Mater. Chem.* **11**, 2482 (2001).
[74] C. Velasco-Santos, A.L. Martinez-Hernandez, and V.M Castano In preparation.
[75] J. N. Coleman, S. Curran, A.B. Dalton, A.P. Davey, B. McCarthy, W. Blau and R.C. Barklie, *Phys. Rev. B.* **58**, R7492 (1998).
[76] C. A. Cooper, D. Ravich, D. Lips, J. Mayer and H.D. Wagner, *Compos. Sci. Technol.* **62** (2002) 1105.
[77] C. Velasco-Santos, A.L. Martínez-Hernández and V.M. Castaño In *"Progress in Nanotechnology Research"* Vol. II Ed. Nova Science Publishers Inc. to be published (2005).
[78] T. W. Ebbesen, *Carbon Nanotubes Preparation and Properties*; CRC Press: Boca Raton, FL, (1997).
[79] C. Velasco-Santos, A.L. Martínez-Hernández and V.M. Castaño. (Ed.) Eugene V. Dirote. In *"Progress in Nanotechnology Research" Vol.* I, 51-78, Ed. Nova Science Publishers Inc. 2004.
[80] Hirsch, *Angew. Chem. Int. Ed.* **41**, 1853 (2002).
[81] Kukovecz, C. H. Kramberger, M. Holznger, H. Kuzmony, J. Schalka, M. Mannsberger and A. Hirsh, *J. Phys, Chem B.*, **106**, 6374 (2002).
[82] J. Zhang, H. Zhou, Q. Qing, Y. Yang, Q. Li, Z. Liu, X. Guo and Z. Pu, *J. Phys. Chem. B.* **107**, 3712 (2003).
[83] L. S. Wojciech, J. A. Libera,. Y. Gogotsi and Y. Masahiro, *J. Solid. State. Chem.* **160**, 184 (2001).
[84] W. Huang, Y. Lin, S. Taylor, J. Gaillard, A. M. Rao and Y-P. Sun, *Nanolett.* **2**, 231 (2002).
[85] W. Z. Li, S. S. Xie and G. Wang, *Science*, **274**, 1701 (1996).
[86] L. Cao, H. Chen, J. Sun, X. Zhang and F. Kong, *J. Phys Chem. B.* **106**, 8971 (2002).

[87] J. Chen, A.M. Rao, S. Lyuksyutov, M.E. Itkis, M. A. Hamon, H. Hu, R.W. Cohn, P.C. Eklund, P. T. Colbert, R.E. Smalley and R.C.Haddon. *J. Phys Chem. B.*, **105**, 2525 (2001).

[88] Velasco-Santos, A.L. Martínez-Hernández, M. Lozada-Cassou, A. Alvarez-Castillo and V.M. Castaño, *Nanotechnology,* **13**, 495 (2002).

[89] S. Qin, D. Qin, W. T. Ford, D. E. Resasco and J. E. Herrera, *Macromolecules,* **37**, 752 (2004).

[90] S. Qin, D. Qin, W. T. Ford, D. E. Resasco, and Jose E. Herrera, *J. Am. Chem. Soc.* **126**, 170 (2004).

[91] Baskaran, J. W. Mays and M. S. Bratcher, *Polymer*, **46**, 5050 (2005).

[92] J. Cui, W.P. Wang, Y.Z. You, C. Liu and P. Wang, *Polymer,* **45** 8717 (2004).

[93] G. Viswanathan, N. Chakrapani, H. Yang, B. Wei, H. Chung, K. Cho, C. Y. Ryu and P. M. Ajayan, *J. Am. Chem. Soc.* **125**, 9258 (2003).

[94] A.L. Martínez-Hernández, C. Velasco-Santos, M. De Icaza and V.M. Castaño, *e-Polymers,* no. 016. (2003), http://www.e-polymers.org

[95] Velasco-Santos, A.L. Martinez-Hernandez, and V.M Castano In preparation.

[96] A.K. Bledzki, S. Reihmane and J. Gassan, *J. Appl. Polym. Sci.* **59**, 1329 (1996)

[97] Y.L. Liu, C.Y. Hsu, M.L. Wong and H.S. Chen, *Nanotechnology*, **14**, 813 (2003).

[98] H.Z. Geng, R. Rosen, B. Zheng, H. Shimoda, L. Fleming, J. Liu and O. Zhou, *Adv. Mater.* **14**, 1387 (2002).

[99] F.H.Gojny, J.Nastalczyk, Z.Roslaniec and K.Schulte. *Chem. Phys Lett.*, **370**, 820 (2003).

[100] H.W. Goh, S.H. Goh, G.Q. Xu, K.P. Pramoda and W.D. Zhang *Chem. Phys Lett.*, **373**, 277 (2003).

[101] F.J. Owens, *Mater. Lett.* In Press (2005).

[102] L. Liu and H. D. Wagner, *Compos. Sci. Technol.* **65**, 1861 (2005).

[103] Zhao, G. Hu, R. Justice, D. W. Schaefer, S. Zhang, M. Yang and C.C. Han, *Polymer*, **46**, 5125 (2005).

[104] S. Bhattacharyya, C. Sinturel, J.P. Salvetat and M.-L. Saboungi. *Appl. Phys. Lett.* **86**, 113104 (2005).

[105] L. Liu, H. Asa, S. Nuriel and H.D. Wagner. *Adv. Func. Mater.* **15**, 975 (2005).

[106] J. W. Cho, J.W. Kim, Y.C. Jung and N. S. Goo. *Macromol. Rapid Comm.* **26**, 412 (2005).

[107] *http://nanotechweb.org/articles/news/4/7/1/1*

INDEX

A

B

C

D

E

F

G

H

I

K

L

M

N

O

P

T

U

V

W

Y